THE NRO AT FIFTY YEARS A BRIEF HISTORY

DR. BRUCE BERKOWITZ WITH MICHAEL SUK

NIMBLE BOOKS LLC: THE AI LAB FOR BOOK-LOVERS
~ FRED ZIMMERMAN, EDITOR ~

Humans and AI making books richer, more diverse, and more surprising.

Publishing Information

(c) 2023 Nimble Books LLC
ISBN: 978-1-60888-263-2

AI-generated Keyword Phrases

National Reconnaissance Office history; reconnaissance satellites development; challenges in developing and deploying reconnaissance systems; NRO role in providing intelligence against hard target; fear of surprise attack; NRO involvement in arms control; reduction of secrecy; launch vehicle problems; NRO support to military operations; importance of technology in NRO operations; future of NRO; NRO role in supporting US security; era of greater openness; common space operations.

PUBLISHER'S NOTE

The National Reconnaissance Office is to the US intelligence community as aircraft carriers are to the Navy. Without the NRO, the payloads that are absolutely crucial to the IC's mission never get there. Without the NRO, the United States is nearly a premodern military power, intelligence-gathering limited to flying patrol planes in international waters. Without the NRO, our appreciation of the strategic situation is as weak as it was during the era of the (non-existent) "missile gap". Yet while there are dozens of unofficial and official histories of the CIA, this official publication from 2011 is the only one in print about NRO. That makes this an essential purchase for the library of anyone with a professional interest in space power or the military history of space.

I am particularly pleased to bring this volume to the public because in a past role I visited the NRO's Center for the Study of Reconnaissance to ask for help on a project I was working on, and the staff there were entirely kind and courteous.

Fred Zimmerman
Ann Arbor, Michigan, USA

ANNOTATIONS

ABSTRACTS

TL;DR (ONE WORD)

History.

EXPLAIN IT TO ME LIKE I'M FIVE YEARS OLD

The National Reconnaissance Office (NRO) is an organization that makes special satellites to spy on other countries. These satellites are used to gather information about what other countries are doing, especially if they might want to harm the United States.

The NRO started a long time ago and had to figure out how to make these special satellites. It was really hard because they needed to make sure the satellites could stay hidden and gather information without getting caught. They also had to worry about other countries attacking the satellites

TL;DR (VANILLA)

The document provides a brief history of the NRO and its development of reconnaissance satellites. It discusses challenges faced, involvement in arms control, launch vehicle problems, and support to military operations. It emphasizes the importance of technology and concludes by discussing the future role of the NRO in supporting US security.

SCIENTIFIC STYLE

This document provides an overview of the National Reconnaissance Office (NRO) and its efforts in developing and deploying reconnaissance satellites. The NRO's role in providing intelligence against challenging targets and the fear of surprise attacks is highlighted. The document also discusses the NRO's involvement in arms control, reduction of secrecy, launch vehicle difficulties, and support to military operations. The significance of technology in the NRO's operations is emphasized. The future of the NRO and its role in supporting US security in an era of increased openness and common space operations is also discussed.

ACTION ITEMS

Continue to invest in research and development of advanced reconnaissance satellite technology to stay ahead of potential adversaries.

Strengthen collaboration with other intelligence agencies and international partners to enhance information sharing and coordination in space operations.

Address launch vehicle problems by working closely with commercial space companies and exploring alternative launch options.

Continue to support military operations by providing timely and accurate intelligence to support decision-making and mission planning.

Explore opportunities for arms control and reduction of secrecy to promote transparency and build trust among nations.

Foster partnerships with academia and industry to leverage their expertise and innovation in developing cutting-edge technologies for reconnaissance purposes.

VIEWPOINTS

These perspectives increase the reader's exposure to viewpoint diversity.

GROUNDS FOR DISSENT

Ethical concerns about surveillance: A member of the organization responsible for this document may have principled reasons to dissent if they believe that the NRO's surveillance activities, enabled by reconnaissance satellites, violate individuals' rights to privacy and civil liberties. They may argue that the NRO's extensive monitoring capabilities go beyond what is necessary for national security and encroach upon citizens' personal lives.

Opposition to military intervention: Another reason for dissent could stem from a belief that military operations supported by the NRO are unjust or unnecessary. This member may argue that investing significant resources in reconnaissance systems and intelligence gathering for military purposes perpetuates a cycle of aggression and conflict. They may advocate for a shift towards diplomacy and peaceful resolution of international disputes rather than relying on military capabilities.

Disagreement with arms control stance: The document mentions the NRO's involvement in arms control efforts, but a dissenting member might hold the view that the organization's approach to arms control is inadequate or misguided. They may believe that the NRO should prioritize disarmament and non-proliferation efforts more forcefully, rather than focusing solely on maintaining a technological edge over potential adversaries.

Concerns about excessive secrecy: While the document briefly touches on the reduction of secrecy within the NRO, a member may dissent if they feel that the organization continues to operate with excessive levels of secrecy. They may argue that transparency and accountability are crucial for ensuring public trust and proper oversight of intelligence agencies, and that a lack thereof undermines democratic principles.

Skepticism towards technological reliance: The emphasis placed on technology throughout the document may be viewed skeptically by a dissenting member who believes that an overreliance on advanced surveillance capabilities can lead to complacency and neglect of other important aspects of national security. They may argue for a more balanced approach that also values human intelligence, diplomatic efforts, and cooperation with international partners.

Overall, a member of the organization responsible for this document might have principled, substantive reasons to dissent based on ethical concerns about surveillance, opposition to military intervention, disagreement with the NRO's arms control stance, concerns about excessive secrecy, and skepticism towards technological reliance. These dissenting views would stem from a different set of values and priorities, reflecting a broader debate within the organization about the appropriate role and actions of the NRO.

RED TEAM CRITIQUE

Overall, the document provides a comprehensive overview of the National Reconnaissance Office (NRO) and its development of reconnaissance satellites. It effectively outlines the challenges faced by the NRO in developing and deploying these systems and highlights their role in providing intelligence against hard targets and protecting against surprise attacks. The discussion on the NRO's involvement in arms control and reduction of secrecy is also informative, as it demonstrates the agency's commitment to transparency and international cooperation.

One strength of the document is its emphasis on the importance of technology in the NRO's operations. It rightly recognizes that technological advancements have played a crucial role in the agency's ability to gather intelligence and support military operations. However, it would have been beneficial to provide more specific examples or case studies to illustrate how technology has been utilized by the NRO.

Additionally, while the document briefly mentions launch vehicle problems, it does not delve into this issue in detail. Given the significance of reliable launch vehicles for successful satellite deployment, it would have been helpful to explore this topic further, including any specific challenges or setbacks the NRO has faced in this regard.

Furthermore, the document could benefit from a more in-depth analysis of the NRO's support to military operations. While it acknowledges the agency's role in this area, it only scratches the surface and fails to provide concrete examples or case studies of the NRO's contributions. Expanding on this topic would enhance the document's credibility and provide a more comprehensive understanding of the NRO's impact.

Finally, the document concludes with a discussion on the future of the NRO and its role in supporting US security in an era of greater openness and common space operations. While this is a relevant and thought-provoking topic, the document falls short in providing any substantive analysis or insights on this matter. It would have been valuable to explore potential challenges and opportunities the NRO may face in adapting to changing circumstances and how it can continue to contribute to US security in a rapidly evolving landscape.

In conclusion, while the document provides a broad overview of the NRO and its development of reconnaissance satellites, there are areas where it could be strengthened. This includes providing more specific examples or case studies, exploring launch vehicle problems in greater detail, delving into the NRO's support to military operations, and offering a more substantive analysis of the agency's future role and challenges. By addressing these shortcomings, the document would become a more comprehensive and insightful resource.

MAGA PERSPECTIVE

The NRO and its development of reconnaissance satellites may seem impressive to some, but it is just another example of how the deep state and the military-industrial complex have been wasting taxpayer money for decades. While they boast about their role in providing intelligence against hard targets, we must ask ourselves, who exactly are these hard targets? It's likely just innocent countries that don't align with the globalist agenda.

Furthermore, the NRO's involvement in arms control is concerning. We should be focused on building up our own military strength, not trying to negotiate with other countries. This is just another way for the globalists to weaken America and compromise our national security.

The document's mention of reducing secrecy is laughable. The government is notorious for its lack of transparency, and it's clear that the NRO is no exception. They claim to support military operations, but how can we trust an agency that operates in secret and keeps the American people in the dark?

And let's not forget about the launch vehicle problems. If the NRO can't even get their satellites into space successfully, how can we rely on them to provide accurate intelligence? It's a waste of time and money.

Finally, the future of the NRO and its role in supporting US security in an era of openness and common space operations is questionable at best. The globalists are working tirelessly to undermine America's sovereignty, and the last thing we need is an agency like the NRO furthering their agenda. It's time to put America first and dismantle this unnecessary and wasteful organization.

Page-by-Page Summaries[1]

2   A brief history of the National Reconnaissance Office, written by Dr. Bruce Berkowitz and Mr. Michael Suk.

3   The Center for the Study of National Reconnaissance is a research body that provides historical context and analysis to the National Reconnaissance Office. Its mission is to study and understand national reconnaissance in order to inform policy and programmatic decisions. Contact information and purchasing details are provided.

4   This page provides an overview of the history and development of the National Reconnaissance Office, including the establishment of reconnaissance satellites, arms control verification, and the challenges faced in the post-Cold War era.

5   N/A

6   The National Reconnaissance Office (NRO) has evolved from a secretive organization to one that can now be more open about its mission. It has played a critical role in analyzing Soviet military forces, arms control agreements, and global operations against terrorists. The challenge for the NRO is to continue supporting US security while adapting to greater public openness and advancements in technology.

7   This monograph provides an account of the establishment, evolution, and accomplishments of the National Reconnaissance Office as it celebrates its fiftieth anniversary. It examines the agency's achievements and future challenges using unclassified and declassified material.

8   This page provides a brief history of the National Reconnaissance Office, highlighting the role of scientists and engineers in developing reconnaissance technology to defend the United States from surprise attacks. It also mentions key individuals who played important roles in the establishment of the organization.

9   The page discusses the importance of technology in national reconnaissance, specifically in response to the challenges posed by the Soviet Union during the Cold War. It highlights the role of high-altitude aircraft, signals intelligence systems, and photoreconnaissance satellites in gathering intelligence and defending the United States. The page emphasizes the ongoing need for technological advancements in addressing future challenges.

10   The page discusses the need for intelligence against a hard target and the fear of surprise attack during the Cold War. It highlights the establishment of the National Reconnaissance Office (NRO) and its role in satellite reconnaissance.

---

[1] Page number *n* in this and the Notable Passages section corresponds to the page labelled in the following facsimile edition as BODY-*n*.

11	During the Cold War, the United States had limited information about the Soviet Union. The Soviets controlled information and were secretive about their military. American leaders turned to high-altitude aircraft for intelligence gathering.
12	American and British intelligence missions, known as SENSINT flights, were conducted over Soviet territory in the 1950s to gather information. These flights were dangerous and yielded limited results. In search of better options, President Eisenhower approved the development of the U-2 reconnaissance aircraft, which utilized new technologies for improved intelligence collection.
13	The page discusses the development of the U-2 aircraft by the CIA and Air Force during the Cold War, as well as the search for alternatives due to increasing Soviet air defense. Options included building a faster aircraft or developing an orbiting satellite.
14	The page discusses the concept of artificial satellites and their potential for communication and military purposes. It highlights the efforts of various American armed services in developing rockets and spacecraft, including the Army's development of Explorer-1 and the Navy's proposal for a multi-service satellite program.
15	The page discusses the development of the Viking launch vehicle and its role in the creation of the first American satellite program. It also mentions the Air Force's development of long-range missiles.
16	The page discusses how the development of reconnaissance satellites was indirectly accelerated by the development of smaller warheads and intercontinental missiles, leading to the establishment of the Western Development Division and the creation of the Atlas, Thor, and Titan missiles. The Air Force's missile and satellite programs were closely linked, and the urgency to develop reconnaissance satellites increased after the Soviet launch of the world's first intercontinental ballistic missile and artificial satellite.
17	NRL's Grab program, derived from the Vanguard program, aimed to develop space systems for electronic intelligence collection. NRL engineer Reid D. Mayo proposed mounting a solid state version of a periscope-mounted radar detector in a satellite. Some Vanguard technicians remained with NRL to form the Satellite Technologies Branch.
18	President Eisenhower approved the Grab program in 1959, which used a satellite to detect radar signals from Soviet air defense systems. The program had two successful missions and three failures. The Poppy program, a larger and more capable satellite, was developed afterwards. The Corona program, which included plans for film scanning and transmission, was also established during this time.
19	N/A
20	The page discusses the development and success of the Corona program, a satellite reconnaissance program that provided accurate information about Soviet missiles during the Cold War. The program was initially intended as a temporary solution but remained in service for twelve

years. The imagery from Corona helped dispel the notion of a "missile gap" and was declassified in 1995.

21   The page discusses the establishment of the National Reconnaissance Office (NRO) and the conflicts and challenges it faced in its relationship with the Secretary of Defense and the Intelligence Community. It also mentions the competition between the Air Force, CIA, and Navy components within the NRO for satellite reconnaissance projects.

22   The page discusses the competition and collaboration between the Air Force, CIA, and Navy in the satellite business, as well as the organizational structure of the NRO.

23   The page discusses the evolution of the National Reconnaissance Office (NRO) and the shifting balance of power between the Department of Defense and the Intelligence Community in terms of planning and authority.

24   The page discusses the relationship between the Defense Department and Intelligence Community in running the NRO, as well as the history of early NRO programs such as Corona and Gambit.

25   The Gambit satellite program, derived from Corona operations, had a slower pace but achieved early success with its first launch in 1963. It completed 14 missions and achieved ground resolutions of better than two feet. By the late 1960s, only three missions failed to produce intelligence.

26   The page discusses the replacement of the Corona satellite system with the Hexagon system, which was larger and more capable. The Hexagon system had multiple film-return buckets and acquired imagery with a resolution of 2-3 feet. It also carried a Mapping Camera System on some flights.

27   The page discusses the Hexagon and Quill missions, which were satellite programs aimed at improving mapping and reconnaissance capabilities. The Hexagon mission was successful but ended with a failed launch, while the Quill mission focused on testing the feasibility of using synthetic aperture radar for imaging.

28   The Quill program successfully collected radar imagery from satellites, but with poor resolution. It was the only early NRO program completed on time and under budget. Satellite reconnaissance became crucial for arms control negotiations between the US and Soviet Union.

29   Arms control treaties influenced NRO programs and led to the gradual reduction of secrecy surrounding satellite reconnaissance. President Nixon chose not to acknowledge satellite capabilities in the SALT treaty, but President Carter later acknowledged them to gain support for SALT II. Information about NRO systems also emerged during espionage trials. The NRO developed electro-optical systems in the 1970s to improve responsiveness.

30   The NRO's imaging systems were too slow to respond to the surprise attack on Israel in 1973. The NRO developed new capabilities, but faced launch vehicle problems and relied on a fragile constellation of satellites. The Space Shuttle was seen as a risky option.

31 The page discusses the challenges faced by the National Reconnaissance Office (NRO) in maintaining its satellite constellation and providing real-time intelligence support during the Gulf War. The NRO developed the Evolved Expendable Launch Vehicle (EELV) to improve launch capabilities, but ground infrastructure limitations continue to be a constraint. President Clinton's directive in 1995 prioritized support to military forces as a top intelligence priority.

32 The NRO faced budget cuts and had to reduce costs by consolidating programs and adopting contracting practices. The Fuhrman Commission recommended consolidating the "Alphabet Programs" and declassifying information. The NRO reorganized into three directorates and later established a fourth directorate for new satellite reconnaissance.

33 The page discusses the declassification of the existence of the NRO and controversies surrounding its funding and headquarters.

34 The NRO failed to properly account for building costs and had excessive carryover funds, damaging their reputation.

35 The NRO accumulated surplus funds due to the extended process of planning, building, and launching satellites. The funds were kept in reserve to ensure a steady supply of critical components and sustain the industrial base. The NRO did not spend the funds on unauthorized programs. The lack of a single accounting system made it difficult to determine the exact amount of accumulated funds. NRO programs operate differently from typical Defense Department weapons acquisition programs.

36 The page discusses the unique challenges faced by the National Reconnaissance Office (NRO) in building satellites and managing budgets. It also highlights the forward funding controversy and the changes implemented to improve financial management.

37 The NRO underwent a significant redesign of its satellite capabilities due to cost reduction pressures and the need to regain its reputation for innovation. The SIGINT component proceeded smoothly, but the IMINT components faced more challenges.

38 The NRO faced challenges with the FIA program, leading to delays and overruns. A commission recommended more funding and prioritization for the NRO, which may have improved the outcome. The NRO now operates a global system of satellites and ground stations for intelligence support.

39 The NRO collects intelligence through satellite imaging and analysis of specific targets. Aging satellites pose a challenge, but the NRO relies on ground stations and international partnerships to support its operations.

40 The NRO plays a crucial role in supporting military operations against terrorist organizations and insurgencies. They collect intelligence using various methods and often work with other intelligence agencies and international partners. The NRO has had to adapt its operations to deal with new threats and technologies used by adversaries.

NOTABLE PASSAGES[2]

6   "In thinking about how far the NRO has come in the past fifty years, the challenge for the reader is to imagine how this national resource can continue to support U.S. security by testing the limits of technology in an era in which the American public expects greater openness and in which space operations have become commonplace. Hopefully, the lessons of scientists, engineers, and intelligence officers who created the NRO will inspire their successors, who will take the organization to even greater achievements."

7   n/a

8   "President Dwight Eisenhower is perhaps most important of all. Driven by his desire to avoid 'another Pearl Harbor,' Eisenhower provided presidential leadership that accelerated overhead reconnaissance efforts and protected those early efforts when failures occurred more frequently than successes."

9   "National reconnaissance systems would not only prove essential in winning the Cold War, but in combating other adversaries faced by the United States. The tides of history set a regular pattern where such forces of intellect will no doubt be required to meet such challenges in the future. Accordingly, space will remain a critical vantage point for ongoing vigilance in defense of the United States."

10   "The greatest fear of U.S. officials in the early years of the Cold War was the potential of a Soviet surprise attack."

11   "The problem was, the United States was largely ignorant about its new adversary. Today, more than two decades after the end of the Cold War, it is hard to appreciate just how little data was available about the Soviet Union at the time. The Soviet government controlled virtually all significant information. It was a major challenge just to find accurate maps of the country, basic economic data, or even a Moscow telephone directory. And the Soviets were especially secretive about their military. There was almost no public data on Soviet forces or weapons systems. Information about Soviet nuclear weapons was nearly nonexistent."

12   "These missions were dangerous, as the Soviet PVO Strany (National Air Defense Forces) regularly attempted to intercept the flights. Several aircraft were downed; others returned damaged. The American and British flyers collected valuable information, but in very limited quantities. Targets like Vladivostok, Shanghai, and Minsk were imaged just a handful of times during the entire life of the program. Targets deeper in denied territory, like Moscow or most Soviet military factories east of the Urals, were not imaged at all."

13   "If uniformed personnel of the armed services of the United States fly over Russia, it is an act of war, legally and I don't want any part of it."

---

[2] Ibid.

14   "An 'artificial satellite' at the correct distance from the earth would make one revolution every 24 hours; i.e., it would remain stationary above the same spot and would be within optical range of nearly half the earth's surface. Three repeater stations, 120 degrees apart in the correct orbit, could give television and microwave coverage to the entire planet." (Clarke, 1945, p. 58)

15   n/a

16   "The 'missile gap' became one of the hottest issues in American politics as legislators such as Sen. Stuart Symington (D-Mo.) and Sen. Henry Jackson (D-Wash.), along with newspaper columnists such as Stewart Alsop, warned that the United States had fallen behind the Soviet Union in strategic nuclear forces. The Sputnik launch boosted support for satellite programs from both Congress and the American public."

17   n/a

18   "The recorded data revealed the location and capabilities of each Soviet radar installation and was used for planning missions to penetrate Soviet airspace in wartime."

19   n/a

20   "Corona imagery showed that the Soviets had far fewer strategic missiles than was thought and dispelled the notion in the early 1960s of a 'missile gap.' For the remainder of the Cold War, satellite IMINT, combined with SIGINT, consistently gave U.S. officials accurate estimates of how many missiles, bombers, and submarines the Soviet Union had at any point in time."

21   "Many scholars and former officials have written about this relationship, often emphasizing the conflict between various components, or between individuals. Make no mistake; this conflict was real. But its actual impact has probably been overstated. In reality, the Air Force, CIA, and Navy programs, though distinctly separate organizations, often worked together, drawing on common technology and support infrastructure."

22   "When the three programs were combined, they had to reconcile their requirement priorities, and this was often a competitive process. This competition was not necessarily bad. At the time, space reconnaissance was the only effective means to monitor Soviet nuclear forces. Competition encouraged multiple solutions to a problem like anticipating the capabilities of a new Soviet weapon system."

23   "The central issue has been whether the Secretary of Defense or the head of the Intelligence Community should have greater say over NRP planning. Some histories (especially those that deal with the NRO's early years) argue that the Defense Department prevailed over the Intelligence Community. In reality the balance has shifted back and forth. Statutes and policies created overlapping authorities."

24   "A certain amount of conflict between the Defense Department and Intelligence Community is an inherent part of the NRO's history, and remains a challenge for running the organization today. It will not go away because it is driven by a budget that is inevitably finite and by

differences in organizational priorities, which are themselves inevitable and which also change over time."

25   "The very first Gambit launch on 12 July 1963 was successful and produced imagery of intelligence targets, something that took 14 Corona launches to achieve."

26   "After years of delays, due as much to politics and organizational tensions as to technological developments, the replacement for Corona, now known as Hexagon (KH-9), flew for the first time on 15 June 1971. By the early 1970s, launcher technology had increased as much as camera technology, so Hexagon, carried by the Titan IIID launcher, was as big as a locomotive and was a much more capable system than Corona."

27   "In April 1960, the U.S. Army unveiled pictures of American cities taken at night and through clouds using a synthetic aperture radar (SAR) system mounted in a small aircraft. This emerging technology was receiving significant interest from people and organizations involved in reconnaissance activities. The Air Force was particularly interested to see if this technology could be used to provide usable post-strike damage assessments without having to wait for appropriate conditions for optical sensors."

28   "Due to the limited scope of the experiment and Maj Bradburn's leadership, the Quill program was the only early NRO program to be completed on time and under budget."

29   "Arms control also triggered a process in which the NRO and its activities gradually became less secret. When the Nixon administration prepared SALT for Senate ratification in 1971, senior officials debated whether to acknowledge that satellite reconnaissance was a 'national technical means' referred to in the treaty. President Nixon decided not to, concerned that countries other than the Soviet Union might object to U.S. surveillance. The Senate ratified the treaty without explicitly discussing the exact definition of national technical means, or the NRO and its capabilities."

30   "The NRO greatly expanded its capabilities during this period. Electro-optical systems introduced when James Plummer served as NRO Director and then fully implemented under his successor, Thomas Reed, were a major breakthrough that revolutionized satellite photoreconnaissance, totally replacing film return systems."

31   "The need for intelligence will grow as next generation weapons enter the inventory. And as the sophistication of weapons increases, deficiencies in intelligence support will proportionally constrain their effectiveness." (House Armed Services Committee, 1991)

32   "During the 1980s and the early 1990s, the NRP had grown annually, even accounting for inflation. With the Cold War over, Democrats and Republicans both supported a 'peace dividend' by cutting defense and intelligence spending. Between Fiscal Year 1990 and 1997 the budget of the National Foreign Intelligence Program declined by 14 percent."

35   "The accumulated funds were a result of several factors. As satellite technology matured, NRO satellites grew larger, became more complex and took longer to build. As noted previously, they also became more reliable and thus often lasted longer, although it was hard to forecast how long. By the 1990s, planning, building, and launching a satellite had become a process that extended over several years, with more uncertainty in knowing exactly when the satellite would be needed on orbit."

36   "In many respects, the practices criticized in the forward funding controversy—a highly compartmented process that minimized administration in favor of success on a tight schedule—were the very practices that had originally justified the NRO's establishment."

37   "Never before in the history of the NRO has the government embarked upon such a significant change in all of its satellite capabilities." (Hill, 2001)

38   "In May 2003, a joint Defense Science Board and Air Force Scientific Advisory Board task force found that FIA was 'significantly underfunded and technically flawed.' In 2005, DNI John Negroponte decided the issue and terminated FIA as it had originally been constituted. The NRO began developing a new strategy for IMINT. In 2009 NRO Director Bruce Carlson described in general terms a new electro-optical system, Next Generation Electo-Optical (NGEO) that will be a lower-risk modular system, capable of being modified in increments over its lifetime."

39   One challenge that the NRO grapples with today is the increasing age of its satellite systems. Currently some NRO satellites are more than 20 years old. This is partly good news because it reflects the improvements in satellite lifetime that the NRO has achieved. The design margins originally needed to meet minimum requirements for reliability have typically allowed a satellite to greatly exceed its planned lifespan. However, it also means that the level of service that the NRO currently provides depends on an aging, and thus increasingly fragile, constellation. (Carlson, 2010)

40   "In these new conflicts, U.S. forces often must find specific individuals—terrorist leaders, financiers, bomb-makers and other 'high value targets' (HVTs)—or specific objects, such as WMD components. Often the NRO has had the only collection capability that could provide the intelligence that U.S. officials and military forces require. To do this, the NRO has had to rethink its operations to deal with these new threats."

41   "Because today's threats can change tactics and methods so quickly, the NRO has put greater emphasis on the ground segment of its systems. Though designing and building a new satellite today can require several years, it is often possible to develop a new data processing system or software tool in a few months to exploit data from the existing satellite constellation. By focusing on the ground segment of a system, NRO can make more frequent modifications and add additional capabilities more easily."

42   "As the users of NRO-derived intelligence grew, it became impractical for the NRO to conceal the basic features of its systems and operations as it had in its early years. However, some experts believed that opening up the NRO had the unintended effect of making it harder to protect truly sensitive capabilities, and this, in turn, made it harder for the NRO to develop the kinds of breakthrough systems that it was known for in its early years."

43   "After fifty years, the challenge for the NRO is to maintain the reliability and contain the costs of its current systems, while at the same time providing the opportunity and challenge that attracts the nation's top minds to imagine new ways to protect American security with overhead reconnaissance."

THE NATIONAL RECONNAISSANCE OFFICE AT 50 YEARS:
A BRIEF HISTORY

SECOND EDITION

THE NATIONAL RECONNAISSANCE OFFICE AT 50 YEARS:
A BRIEF HISTORY

by Dr. Bruce Berkowitz
with Mr. Michael Suk
Center for the Study of National Reconnaissance
National Reconnaissance Office
Chantilly, Virginia
July 2018
Second Edition

CENTER FOR THE STUDY OF
NATIONAL RECONNAISSANCE

Center for the Study of National Reconnaissance (CSNR)

The Center for the Study of National Reconnaissance is an independent National Reconnaissance Office (NRO) research body reporting to the NRO Deputy Director, Business Plans and Operations. Its primary objective is to ensure that the NRO leadership has the analytic framework and historical context to make effective policy and programmatic decisions. The CSNR accomplishes its mission by promoting the study, dialogue, and understanding of the discipline, practice, and history of national reconnaissance. The Center studies the past, analyzes the present, and searches for lessons-learned.

Contact Information: To contact the Center for the Study of National Reconnaissance, please phone us at 703-488-4733 or email us at csnr@nro.mil

To Obtain Copies: Government personnel can obtain additional printed copies directly from CSNR. Other requestors can purchase printed copies by contacting:

Government Printing Office
732 North Capitol Street, NW
Washington, DC 20401-0001
http://www.gpo.gov

Published by
National Reconnaissance Office
Center for the Study of National Reconnaissance
14675 Lee Road
Chantilly, Virginia 20151-1715

Printed in the United States of America
ISBN: 978-1-937219-17-8

Contents

Foreword

The history of the National Reconnaissance Office is a story of how opportunity, necessity, and determination converged to produce an intelligence organization unlike any that had come before. In the late 1950s, rocket and sensor technologies were just reaching a level of maturity so that, if pushed to the limit, they could assist the United States in facing the most challenging national security problem of the age: how to analyze Soviet military forces and avert a potential nuclear war.

After providing the hard data that made it possible to understand and deter the Soviet Union, NRO systems later became the primary means that made possible the arms control agreements that defused U.S.–Soviet tensions. After the collapse of the Soviet Union, NRO systems became ever more integrated into U.S. military capabilities, playing a critical role in the Gulf Wars, peacekeeping operations, and most recently, global operations against terrorists.

In retrospect, it seems remarkable that even as the United States was achieving its goal of putting a man on the moon, there was an equally ambitious and technologically challenging American space program proceeding along a parallel path—but in strictest secrecy. Indeed, it was not until 1978 that a President acknowledged the basic fact that the United States carried out reconnaissance from space, and not until 1992 that the government acknowledged the NRO's existence.

Until recently, it would have been impossible to publish an official, authoritative, unclassified history of the NRO. The fact that we can tell the history of this second space program today shows how much the NRO has evolved. Originally the NRO and its mission were totally unacknowledged, first to protect the source and method; and second, in deference to the sensitivity that some countries might have to U.S. satellites orbiting over their territory. Today we take such activities for granted, and the NRO and its mission can be much more open and focus its measures for secrecy on those areas in which the organization is developing technologies that exceed the public's imagination and the expectations of our adversaries.

In thinking about how far the NRO has come in the past fifty years, the challenge for the reader is to imagine how this national resource can continue to support U.S. security by testing the limits of technology in an era in which the American public expects greater openness and in which space operations have become commonplace. Hopefully, the lessons of scientists, engineers, and intelligence officers who created the NRO will inspire their successors, who will take the organization to even greater achievements.

Robert A. McDonald, Ph.D.
Director, Center for the Study of National Reconnaissance
National Reconnaissance Office
September 2011

Preface

The objective of this monograph is to provide an account of the establishment, evolution, and accomplishments of the National Reconnaissance Office as it marks its fiftieth anniversary. It draws on the extensive body of unclassified and declassified material about the NRO that is available today to examine the agency's accomplishments and the challenges it will face in the future. This monograph would not have been possible without the support of others. In particular, thanks are due to Robert A. McDonald, James D. Outzen, and Jimmie D. Hill for comments on the manuscript and to Karen Early for guiding it through production.

Bruce Berkowitz, Ph.D.
Center for the Study of National Reconnaissance
National Reconnaissance Office
September 2011

Introduction

Nearly twenty years after the Japanese attacked Pearl Harbor, one of the final chapters of World War II history opened when acting CIA Director Gen Charles Cabell established the National Reconnaissance Office by concurring with Deputy Secretary of Defense Roswell Gilpatric's 6 September 1961 memorandum. The ghosts of Pearl Harbor loomed large indeed as teams of extraordinary scientists developed high-altitude and satellite technology, hoping to assure that the United States would never again face a devastating, surprise attack.

Three forces molded the subsequent chapters in the National Reconnaissance Office's history: brilliant scientists and engineers, stunning reconnaissance technology, and hard intelligence challenges. In his brief history of the National Reconnaissance Office, Bruce Berkowitz presents unclassified glimpses of the scientists and engineers, reconnaissance technologies, and intelligence issues that drove the development of the National Reconnaissance Office and its efforts to defend the nation during the last fifty years of air and space advances.

A highly talented group of individuals with diverse backgrounds played important roles in the establishment of the National Reconnaissance Office. These individuals include Dr. James Killian, science advisor to President Eisenhower and President of the Massachusetts Institute of Technology, who provided critical support for national reconnaissance systems. Dr. Edwin "Din" Land, inventor of instant photography and President of the Polaroid Corporation, became an influential advocate for the use of new technology to solve intelligence puzzles. Dr. Richard Bissell, the talented Marshall Plan administrator, applied those management skills to develop two early successful reconnaissance programs—the U-2 high-altitude spy plane and the nation's first photoreconnaissance satellite (Corona). Dr. Joseph Charyk, who would later lead one of the nation's largest commercial satellite companies, provided early and essential leadership for the nation's first overhead reconnaissance organization. President Dwight Eisenhower is perhaps most important of all. Driven by his desire to avoid "another Pearl Harbor," Eisenhower provided presidential leadership that accelerated overhead reconnaissance efforts and protected those early efforts when failures occurred more frequently than successes.

Developing national reconnaissance technology provided an unprecedented vantage point for the United States, but also posed unprecedented technological challenges. One of the most significant early challenges included developing a plane that could fly in the thin atmosphere of high altitudes. Another significant challenge was the launch of large objects into space on a regular basis without failure. Once in orbit, technological challenges existed for retrieving information—pictures and signals— from space. Space is a harsh environment, and space technology must persist in that environment. Consequently, scientists and engineers developed new materials: film, lenses, antennas, and other components to survive in space. The U-2, the nation's first high-altitude aircraft; Grab, the first signals intelligence system; and Corona, the first photoreconnaissance satellite, exemplify technology's successful response to these challenges. For these systems, existing technology was reshaped to operate in space. Technological breakthroughs allowed the United States to gather signals and photograph adversaries from high altitudes and the far reaches of space. Early efforts brought disappointment as often as success, but as technology matured, overhead reconnaissance proved to be a reliable asset in the defense of the nation.

Leaders of the United States struggled with difficult national security questions when the Soviet Union emerged from the aftermath of World War II as the nation's most formidable foe. With its August 1949 nuclear test, the Soviet Union became the world's second nuclear power. From that point forward, the United States faced questions about Soviet capabilities for delivering nuclear weapons in an attack on the United States. Although Soviet leadership intentions remained largely opaque, Soviet capabilities became more transparent as a result of intelligence gained from national reconnaissance systems. By the mid-1950s, controversies arose whether long-range bombers in the Soviet arsenal exceeded the number of those in the United States. The U-2 would largely settle that question. Shortly thereafter, questions arose about the Soviet's capabilities to deliver nuclear weapons on intercontinental ballistic missiles. Corona and subsequent photoreconnaissance and signals intelligence satellite systems, such as Gambit and Hexagon, would help settle those questions. As the United States gained more insight into Soviet military capabilities, new intelligence challenges arose in the 1970s when arms limitation treaties required verification. Scientists and engineers would once again leverage technology to answer these intelligence challenges from space.

Although this is an unclassified history of national reconnaissance, and consequently an incomplete history, it nonetheless provides glimpses into the force of human intellect in shaping technology to address difficult questions. National reconnaissance systems would not only prove essential in winning the Cold War, but in combating other adversaries faced by the United States. The tides of history set a regular pattern where such forces of intellect will no doubt be required to meet such challenges in the future. Accordingly, space will remain a critical vantage point for ongoing vigilance in defense of the United States.

James D. Outzen, Ph.D.
NRO Historian, Center for the Study of National Reconnaissance
National Reconnaissance Office
August 2014

The Need for Intelligence Against a Hard Target

The Fear of Surprise Attack

On 25 August 1960, President Eisenhower greeted several of his top science advisors in the Oval Office with Director of Central Intelligence, Allen Dulles, just before the President was scheduled to attend a meeting of the National Security Council. One of the scientists, Polaroid Corporation's CEO Edwin H. "Din" Land, unrolled a spool of film across the floor.

Land said, "Here are your pictures, Mr. President." The film was from Corona 14, the first successful satellite photoreconnaissance mission, which had flown the week before. Corona 14 had captured images of airfields and other military installations in the Soviet Union. President Eisenhower had approved the project two and a half years before. (McDonald, 2002, p. 34)

Land and another advisor attending the meeting, James R. Killian, Jr., President of the Massachusetts Institute of Technology, had championed Corona through its difficult development. At the time, the United States had several satellite reconnaissance programs underway. The Air Force and Central Intelligence Agency were both developing imagery intelligence, or "IMINT" systems. The Navy had already orbited the first signals intelligence, or "SIGINT" satellite in June.

The experience of Corona led Land and Killian to conclude that a new organization was needed for overhead reconnaissance that minimized bureaucracy. They also believed it should be civilian, and classified. Although the Soviets had established the precedent for passing over national borders in orbit with Sputnik 1, overflight was still sensitive, and the prospect of a military spacecraft openly passing over Soviet territory might have been provocative.

Land and Killian had proposed a new national office responsible for the design, acquisition, and operation of reconnaissance satellites. President Eisenhower had agreed, saying that he regretted not having made the decision two years earlier, when he had approved the Corona program. The group then went to an adjoining room for the NSC meeting, where the President gave his formal approval for the establishment of the National Reconnaissance Office. (McElheney, 1999; Hall and Laurie, 1999)

The NRO can trace its heritage to World War II, when U.S. forces used aircraft to collect imagery and signals intelligence to plan military operations against Germany and Japan. As the Cold War heated up, U.S. officials discovered that overhead reconnaissance was one of the few options available to discover basic facts about the military and industrial capabilities of the new opponent the nation faced, the Soviet Union.

The greatest fear of U.S. officials in the early years of the Cold War was the potential of a Soviet surprise attack. In February 1946, Joseph Stalin had declared that war

Figure 1: Original National Reconnaissance Staff, 1961

with the West was inevitable; the next few years were marked by espionage cases involving Soviet spies, Soviet expansion into Eastern Europe, the Berlin blockade, and the fall of China to communism. Pearl Harbor was still a fresh memory; these fears grew even more after the Soviet Union tested its first atomic bomb on 29 August 1949. The problem was, the United States was largely ignorant about its new adversary.

Today, more than two decades after the end of the Cold War, it is hard to appreciate just how little data was available about the Soviet Union at the time. The Soviet government controlled virtually all significant information. It was a major challenge just to find accurate maps of the country, basic economic data, or even a Moscow telephone directory. And the Soviets were especially secretive about their military. There was almost no public data on Soviet forces or weapons systems. Information about Soviet nuclear weapons was nearly nonexistent.

To make matters worse, the Soviet Union was a difficult target for traditional human intelligence operations. The Soviet government limited travel by foreigners, monitored visitors closely, and discouraged their interaction with Soviet citizens. All of this made the Soviet Union an exceptionally challenging environment for recruiting and running assets.

With few alternatives, American leaders turned to high-altitude aircraft flying near—and sometimes over—Soviet territory. On 4 December 1950, following the invasion of South Korea, British Prime Minister Clement Atlee met with President Harry Truman in Washington. Some historians believe that it was at this meeting

BODY-11

that the two leaders, fearing imminent war with the Soviet Union, agreed to cooperate in overflights of Soviet territory.[1]

That month, President Truman approved two flights over Eastern Siberia from Alaska, and the Air Force pulled one of the first of its new B-47 bombers from the production line to modify it for reconnaissance missions. The program expanded over the next several years, with missions flying out of bases in Alaska, Greenland, Britain, and West Germany. These "SENSINT" flights continued through 1955 under the Eisenhower administration.

These missions were dangerous, as the Soviet PVO Strany (National Air Defense Forces) regularly attempted to intercept the flights. Several aircraft were downed; others returned damaged. The American and British flyers collected valuable information, but in very limited quantities. Targets like Vladivostok, Shanghai, and Minsk were imaged just a handful of times during the entire life of the program. Targets deeper in denied territory, like Moscow or most Soviet military factories east of the Urals, were not imaged at all.

The Air Force also attempted to image Soviet targets with high-altitude balloons in a series of activities that eventually evolved into the Genetrix program. It launched the first of 516 balloons on 10 January 1956, but President Eisenhower cancelled the program just a month later, after the Soviets downed one of the balloons, put its payload on display, and issued an official protest of the American violation of its airspace. (Welzenbach, 1986)

Experts and Technology Step In

Facing this intelligence vacuum, American leaders looked for new options. President Eisenhower and a small circle of scientists and industry advisors became personally immersed in solving the problem. As it happened, new technologies—high-altitude aircraft, long-range rockets, satellites, and sensors—were just beginning to emerge that offered opportunities for intelligence collection far beyond anything that had been achieved before.

On 26 July 1954, President Eisenhower appointed a "Technical Capabilities Panel (TCP)," chaired by Killian to study options to deal with the threat. The TCP's Intelligence Projects Committee, chaired by Land, recommended that the government proceed with a previously offered plan by the Lockheed Corporation to build a reconnaissance aircraft specifically designed to fly above Soviet air defenses. (Pedlow and Welzenbach, 1998, p. 27)

This aircraft became the U-2, which President Eisenhower approved in November 1954. To build the aircraft quickly, the panel recommended using the CIA's special authorities to use "unvouchered funds" and streamlined contracting. This was

---

[1] See, in particular, Hall and Laurie, 2003. The two scholars report that they could not find documentation of the agreement, noting the extreme sensitivity of the proposed operations, but argue that this was when it was made on the basis of later events and circumstantial evidence.

Figure 2: The U-2 Aircraft

a significant step, bringing the CIA into the development of large-scale technical collection systems for the first time. Another reason for using the CIA was to avoid provoking Soviet leaders during an especially tense period of the Cold War. President Eisenhower believed that the overflight had to be conducted as a classified operation. "If uniformed personnel of the armed services of the United States fly over Russia," he wrote, "it is an act of war, legally and I don't want any part of it."

A joint Air Force-CIA team developed the U-2 with Lockheed under Project Aquatone. Richard Bissell, the Special Assistant for Planning and Coordination at the CIA, directed the project with his deputy director, Osmund J. Ritland, then a colonel in the Air Force. Lockheed built the first aircraft in eight months. The first mission was flown on 4 July 1956. A total of 24 successful missions were completed, until 1 May 1960 when CIA pilot Francis Gary Powers was shot down near Sverdlovsk (now Yekaterinburg) by an SA-2 surface-to-air missile.

Early in the U-2 program it became clear that the improving Soviet air defense system would make U-2 missions over the Soviet Union too risky. U.S. officials searched for new alternatives. One was to build a faster, even higher-flying aircraft with a smaller radar signature. Bissell assembled an expert committee to investigate a follow-on to the aircraft. The committee was chaired by Land and met from 1957 to 1959 to review proposals from several aerospace contractors. (Robarge, 2009)

Lockheed, led by its chief engineer Clarence "Kelly" Johnson, proposed a series of concepts it called "Archangel," a play on the company's original name for the U-2, "Angel." When the twelfth concept was adopted, the aircraft became known as the A-12, or by its CIA project name, Oxcart. As with the U-2, the Air Force was a partner throughout the project, providing pilots, training, bases, and logistical support. The Air Force also later developed its own two-seat version of the A-12, the SR-71.

The other option for replacing the U-2 was to develop an orbiting satellite. Science fiction writer Arthur C. Clarke may have been one of the first to propose the basic idea of a satellite and its varied uses. Even before World War II was over, Clarke speculated how one might use the German V-2 as a satellite launch vehicle. In the February 1945 edition of *Wireless World* Clarke wrote,

4

A rocket which can reach a speed of 8 km/sec parallel to the earth's surface would continue to circle it forever in a closed orbit; it would become an 'artificial satellite' ... It would thus be possible to have a hundred-weight of instruments circling the earth perpetually outside the limits of the atmosphere and broadcasting information as long as the batteries lasted. Since the rocket would be in brilliant sunshine for half the time, the operating period might be indefinitely prolonged by the use of thermocouples and photo-electric elements ... .

Clarke anticipated the idea of using geosynchronous satellites for receiving and re-transmitting radio signals from space—the basic concept for both a communications satellite and a satellite for collecting SIGINT. He observed that

An 'artificial satellite' at the correct distance from the earth would make one revolution every 24 hours; i.e., it would remain stationary above the same spot and would be within optical range of nearly half the earth's surface. Three repeater stations, 120 degrees apart in the correct orbit, could give television and microwave coverage to the entire planet. (Clarke, 1945, p. 58)

All of the American armed services saw the potential for using space for military purposes and began programs to develop rockets and spacecraft. But in the lean post-war years, these were just shoestring efforts.

The Army, working with German scientists who had surrendered at the end of the war, concentrated on missiles to supplement traditional artillery. The Army's efforts eventually led to Explorer-1, the first successful American satellite, and later evolved into the nucleus of the National Aeronautics and Space Administration's (NASA) Marshall Space Flight Center, which developed the Saturn moon rocket. NRO and NASA have often worked together. In the 1960s for example, NASA used technology that NRO had developed for earth imaging systems in its Lunar Orbiter camera, and the NRO used NASA's Space Shuttle to launch satellites.

The Navy was also developing space technology. The Naval Research Laboratory (NRL) had conducted upper atmosphere experiments since 1946, using captured German V-2s as "sounding rockets" to carry telemetered instrumentation to altitudes of more than one hundred miles. A space program was the next logical step; in October 1945 the Navy's Bureau of Aeronautics had proposed a satellite program. The following year, the Navy floated the idea of a multi-service effort.

The Army Air Forces, soon to become a separate service, had also experimented with captured V-2s and was considering a satellite program. To prepare for a meeting with his Navy counterparts to discuss their proposal, Maj Gen Curtis LeMay directed the RAND Corporation to develop a feasibility study. The report, *Preliminary Design of an Experimental World-Circling Spaceship,* was an engineering analysis for a generic launch vehicle and satellite. It speculated that such a system might serve

Figure 3: The SR-71 Aircraft

as a communications or a scientific research platform. However, it also mentioned in a single paragraph that such a "satellite offers an observation aircraft that cannot be brought down by an enemy who has not mastered similar techniques." (Douglas Aircraft Company, 1946)

Although the newly-created Air Force rejected the Navy's proposal for a multi-service space program, NRL continued its upper atmosphere research. In 1949, as the supply of V-2s ran out, NRL developed the Viking, the first large launch vehicle combining the familiar elements we recognize today: cylindrical cross section, monocoque construction, gimbaled engines, and a separable payload section. Some Navy personnel who worked on the Viking went on to develop the Delta launch vehicle at NASA. This rocket evolved into today's Delta-IV, a mainstay launch vehicle for the NRO.

The Viking also led to the first American satellite program. In 1954 the International Council of Scientific Unions, an association of national science organizations, designated 1957 as the "International Geophysical Year," or IGY, when countries would carry out a coordinated program of earth science studies. The organizing committee for the IGY proposed that countries might develop a scientific satellite as part of the effort. On 26 May 1955 the Eisenhower administration accepted the proposal in the first U.S. national space policy, and selected NRL to lead Vanguard, a program to develop a research satellite and a Viking-based orbital launch vehicle.

At the same time, the Air Force began the first steps in developing long-range missiles. A June 1953 Air Force Scientific Advisory Board (AFSAB) report had concluded that it was possible to build small, lightweight nuclear warheads.[2] This recommendation

---

[2]   See Greer, K.E. (Spring 1973), "Corona," *Studies in Intelligence,* reprinted in Ruffner, K.C., Ed. (1995) *Corona: America's First Satellite Program.* Washington, DC: Central Intelligence Agency, pp 3-39. Although no official record of the meeting seems to exist, the Air Force Vice Chief of Staff, Gen T.D. White, referred to it in a letter dated 8 June 1953 to Theodore von Karman, cited in Jacob Neufeld, Center for Air Force History, "Technology Push," www.history.mil/colloquia/cch9c.html

BODY-15

indirectly accelerated the development of reconnaissance satellites. At the time the hardest technical challenge in developing a satellite was not the spacecraft, but the launch vehicle. Smaller warheads made nuclear-armed, intercontinental missiles feasible, and these missiles provided the basis for orbital launch vehicles. (Greer, 1973)

The AFSAB recommendation led to the decision by the Air Force on 1 July 1954 to establish the Western Development Division (WDD) under the command of Brig Gen Bernard Schriever and assign it the responsibility for long-range ballistic missile development. The Air Force contracted with Convair to build the Atlas, the first operational American intercontinental ballistic missile (ICBM), which became the highest priority U.S. defense program. It also started development of the Thor, a smaller intermediate-range missile built by Douglas Aircraft Corporation, and the Titan, a much larger missile built by the Martin Company, the contractor on the Navy's Viking. All three were later used as launch vehicles for NRO satellites.

From the beginning, Air Force missile and satellite programs were closely linked. With the availability of a launch vehicle imminent, the Air Staff issued an operational requirement for a means to provide continuous surveillance of an enemy's war-making capability. Under this requirement, the Air Research and Development Command (which had taken over the RAND satellite studies in 1953) approved the WS-117L program.[3] From 1954 to 1957, the WS-117L developed studies for reconnaissance satellites based on Lockheed's Agena upper stage and incorporating a variety of IMINT and SIGINT payloads. On 2 April 1956 the Air Force completed its development plan for a reconnaissance satellite.

Even so, WS-117L languished until the Soviet launch of the world's first intercontinental ballistic missile in August 1957, and a month later, Sputnik-1, the first artificial satellite. Sputnik created a new sense of urgency. The "missile gap" became one of the hottest issues in American politics as legislators such as Sen. Stuart Symington (D-Mo.) and Sen. Henry Jackson (D-Wash.), along with newspaper columnists such as Stewart Alsop, warned that the United States had fallen behind the Soviet Union in strategic nuclear forces. The Sputnik launch boosted support for satellite programs from both Congress and the American public.

The First Reconnaissance Satellites

By 1958 there were two satellite efforts within the Defense Department. One, the Vanguard research satellite program at the Naval Research Laboratory, would provide the basis for Grab, the first SIGINT satellite. The other, the Air Force's WS-117L military system, would provide the basis for two later programs: Corona, developed by the CIA as the first IMINT satellite; and Samos, a satellite program combining SIGINT and IMINT that provided the foundation for the Air Force's space

---

3   See Hall, R.C. (Fall 1963), "Early U.S. Satellite Proposals," *Technology and Culture*, pp. 410-434, cited in Haines (1996).

Figure 4: Reid D. Mayo

Figure 5: Grab, Piggybacked Atop TIROS

reconnaissance program. These three programs, together with the CIA's A-12 and Air Force's SR-71 reconnaissance aircraft programs, were the building blocks of what would become the National Reconnaissance Office.

The Beginning of SIGINT Collection From Space

Grab evolved directly from NRL's Vanguard program. Laboratory officials had discussed plans for using technology from Vanguard to develop space systems for electronic intelligence collection at least as early as 1957, along with other military missions such as communications, navigation, and scientific measurements of atmospheric and space phenomenon that might affect military operations.

NRL had previously developed crystal video technologies deployed on submarines to intercept and analyze Soviet radar sites. Passing time as he was stuck in a Pennsylvania snowstorm in March 1958, Reid D. Mayo, an NRL engineer, came up with the idea of mounting a solid state version of the periscope-mounted radar detector in a Vanguard-like satellite. Returning to Washington, he proposed the idea to Howard Lorenzen, chief of the NRL electronic countermeasures branch. The lab began developing the project for the Director of Naval Intelligence. (McDonald and Moreno, 2005)

When President Eisenhower established the National Aeronautics and Space Administration on 1 October 1958, most NRL technicians working on Vanguard transferred to what is today's Goddard Space Flight Center. However, Robert Morris Page, NRL's Director of Research, believed the Navy still needed a space engineering capability. Some Vanguard technicians remained with NRL, forming the Satellite Technologies Branch, which evolved into today's Naval Center for Space Technology. (van Keuren, 1987)

President Eisenhower approved Mayo's program, by then designated "Tattletale," on 24 August 1959. Grab ("Galactic Radiation and Background," a name referring to the satellite's cover as a research project measuring radiation in space) was equipped with both scientific instruments and a receiver that detected pulsed-radar signals emitted from Soviet air defense systems.

The Air Force launched Grab 1 on a Thor Able Star vehicle on 22 June 1960, becoming the world's first successful reconnaissance satellite. Launched into a 600 mile high, 63.4 degree inclination orbit, Grab passed over the entire territory of the Soviet Union. As the satellite detected signals from air defense radars, it transmitted a corresponding signal to a ground control "hut" that was within line of sight of the satellite, but in friendly territory. The signals were recorded on tapes, which were sent back to NRL, and ultimately, the National Security Agency and the Air Force Strategic Air Command for analysis. The recorded data revealed the location and capabilities of each Soviet radar installation and was used for planning missions to penetrate Soviet airspace in wartime.

The Grab program ended with two successful missions and three failures. NRL went on to develop Poppy, a larger, more capable satellite, first launched on a Thor Agena-D vehicle on 13 December 1962. A total of seven Poppy missions were launched, the last one reaching orbit on 14 December 1971. Grab remained secret to the public until 1998, when its existence was declassified as part of the commemoration of NRL's 75th anniversary. Poppy was declassified in 2004. Both systems proved essential in tracking Soviet air defense radar capabilities.

Corona: The First IMINT Satellite

In January 1956 President Eisenhower established the President's Board of Consultants on Foreign Intelligence Activities (PBCFIA), a predecessor of today's President's Intelligence Advisory Board. It included members of the TCP, including Killian, who chaired the board, and Land. By this time, the Air Force's WS-117L project included plans for a SIGINT payload, a payload in which film from an IMINT payload would be scanned and transmitted to ground stations, and a payload in which film would be returned to earth via a reentry capsule.

Killian and Land doubted it was possible to develop a scanning and readout imagery system with the needed resolution or data volume using the technology available at the time. They were also concerned with the pace of the program, which they believed was too slow to meet the threat presented by the Soviets. These issues came to a head in October 1957 when the PBCFIA issued its semi-annual report to the President, emphasizing the need for a faster approach that would provide an interim capability.

In December 1957 the two men met at the White House with Bissell; the President's staff assistant, Brig Gen Andrew Goodpaster; and Maj Gen Bernard Schriever, head of the Air Force's ICBM and WS-117L projects. They agreed that the best course of action was to break up the WS-117L project and concentrate on the film recovery

Figure 6: Engineers Ready Corona

approach as a separate "crash" effort. President Eisenhower approved this new program, Corona, on 7 February 1958. The downing of Powers' U-2 two years later would provide yet another push for urgency in the Corona program.

The media had linked WS-117L to satellite reconnaissance shortly after the Sputnik launch, so President Eisenhower directed the Air Force to ostensibly kill the film recovery segment, and then secretly transfer it to the CIA. As with the U-2, using the CIA allowed the government to use the agency's special contracting authorities and keep the effort classified, which was critical because of the continuing political sensitivity of overhead reconnaissance. Bissell was appointed to manage the program, and Ritland again joined him as his deputy, along with other members of the U-2 team. (Greer, 1973)

Corona, a complex system using new technology, proved challenging to develop. Twelve missions failed due to launch vehicle, satellite, or recovery system malfunctions. A thirteenth spacecraft was successful, but as a test vehicle, carried no film. The first fully successful mission was launched on 18 August 1960 and recovered the next day. The first mission returned with 3,000 feet of film (more than the entire U-2 program up to then), imaging 1.65 million square miles of Soviet territory.

Corona was originally intended as a stopgap until more capable systems entered service. But the program proved so productive that it was kept in service for almost twelve years and 145 missions, the last being launched on 25 May 1972. The NRO operated several different versions of Corona during the program's lifetime, introducing different camera systems and making incremental improvements. The earliest missions produced imagery with a ground resolution of 40 feet, using the KH-1 camera (KH denoted Keyhole, the name of the program's security system). Later cameras, the KH-2 and KH-3, improved resolution to 10 feet. The KH-4, the final Corona camera system, ultimately produced imagery with 5-7 foot resolution. NRO also provided Corona with multiple film return capsules that extended the film capacity of each mission. (Ruffner, 1995, pp. xiv-xv)

Corona imagery showed that the Soviets had far fewer strategic missiles than was thought and dispelled the notion in the early 1960s of a "missile gap." For the remainder of the Cold War, satellite IMINT, combined with SIGINT, consistently gave U.S. officials accurate estimates of how many missiles, bombers, and submarines the Soviet Union had at any point in time. The Corona program, along with most of its imagery, was declassified in 1995 by an executive order from President Clinton.

Establishment of the National Reconnaissance Office

When President Kennedy entered office in 1961, his Secretary of Defense, Robert McNamara, a management expert who came to the government from Ford Motor Company, generally wanted to streamline the Defense Department's structure by consolidating organizations with like functions; the Defense Intelligence Agency and the Defense Supply Agency were examples. Because space reconnaissance activities

Figure 7: Dr. Joseph V. Charyk

Figure 8: Dr. Richard M. Bissell, Jr.

were already under a civilian chief in the Department of the Air Force, it was natural for McNamara to form a consolidated NRO around that office. So on 6 September 1961, Secretary McNamara and Director of Central Intelligence Allen Dulles agreed to establish the NRO, jointly managed by Bissell and Under Secretary of the Air Force Joseph V. Charyk. The new office would be responsible for the newly-created National Reconnaissance Program (NRP), which subsumed both Corona and Samos.

The relationship of the NRO to the Secretary of Defense and to the head of the Intelligence Community has evolved throughout the organization's existence. Many scholars and former officials have written about this relationship, often emphasizing the conflict between various components, or between individuals. Make no mistake; this conflict was real. But its actual impact has probably been overstated. In reality, the Air Force, CIA, and Navy programs, though distinctly separate organizations, often worked together, drawing on common technology and support infrastructure. Contractors often worked for more than one organization. Nevertheless, putting the CIA's space system activities into a Defense Department organization did create turf issues and chain of command questions. Should the Secretary of Defense control a CIA program as part of the NRO? Making the NRO part of the Intelligence Community created even more issues. Should the Director of Central Intelligence control a Defense Department agency as part of the Intelligence Community? This issue continues even today, with the Director of National Intelligence (DNI) filling the role that the DCI formerly played.

Because there was always a finite budget for satellite reconnaissance, the Air Force, CIA, and Navy components within NRO inevitably competed when an opportunity emerged to build a new kind of system. All three organizations were hungry for a new challenge and the opportunity to contribute to the nation's security. This competition became more intense during periods of declining defense spending. And in addition to all of this, the Air Force, CIA, and Navy each had different reasons for being in the satellite business, and thus different priorities. The Air Force was most concerned with supporting U.S. strategic nuclear force targeting, the CIA with intelligence estimates of the Soviet Union, and the Navy with supporting fleet operations. When the three programs were combined, they had to reconcile their requirement priorities, and this was often a competitive process.

This competition was not necessarily bad. At the time, space reconnaissance was the only effective means to monitor Soviet nuclear forces. Competition encouraged multiple solutions to a problem like anticipating the capabilities of a new Soviet weapon system. In one case, the Air Force proposed a concept to collect electronic and communications intelligence to assess new Soviet force developments, while the CIA proposed a different concept based on collecting instrumentation data from Soviet tests. One offered a lower risk of failure but also a lower payoff; while the other promised a high payoff, but also higher risks. The decision of which approach to use in any particular case was likely to go to the White House, and in this case, the President decided to proceed with both concepts, each of which proved productive. Bissell resigned from the CIA in February 1962 following the failed invasion of the Bay of Pigs; Dulles had resigned the previous November, being replaced by John McCone. Bissell's departure left Charyk the sole director of the NRO. A second agreement between DOD and CIA, signed on 2 May 1962 formalized this fact by establishing a single director for the NRO, who would be a Defense Department civilian designated by the Secretary of Defense with the agreement of the DCI. This agreement also added the Navy's Grab program to the NRP.

Two months later on 23 July 1962, Charyk, now the sole director, established the basic organizational structure that the NRO would keep for the next thirty years:

- Program A, the Air Force satellite reconnaissance program based in El Segundo, California;

- Program B, the CIA satellite reconnaissance program based in the northern Virginia suburbs of Washington;

- Program C, the Navy program, comprised of personnel from NRL, the Navy Security Group, and NSA, based in southeast Washington, DC;

- Program D, the Air Force and CIA aerial reconnaissance program, comprising all national assets, including the U-2 and A-12/SR-71 programs. (This program was dissolved, and its assets were transferred to the Air Force when the CIA's A-12s were deactivated on 1 October 1974).

This structure reflected the fact that the NRO was, in practice, a highly decentralized federation of several existing programs. The NRO Director acted as a "corporate CEO," supported by a small staff, making major decisions on budgets, policy, and program starts. Most of the action involved in planning and building satellites took place in the "alphabet programs." Geographically separated and linked to its own parent organization, each developed its own culture. Organizational boundaries were also reinforced by security regulations, because in the early years of the NRO, satellite programs were more strictly compartmented from each other. As a result, most personnel probably thought of themselves as Air Force, CIA, or Navy as much— or more—than as members of the NRO.

McCone reportedly regretted the CIA's diminished role in the second agreement and pressed for a third agreement, dated 13 March 1963, shortly after the second NRO Director, Brockway McMillan, took office. (Hall, 2002) This agreement designated the Secretary of Defense as the Executive Agent for the NRP, and established the NRO as an agency within the Defense Department. But it also created an NRO Deputy Director, who would be a CIA official. The Director would be appointed by the Secretary of Defense with the agreement of the DCI, and the Deputy Director would be appointed by the DCI with the agreement of the Secretary of Defense. The Director was to report directly to the Secretary of Defense, while keeping the DCI informed.

A fourth agreement was signed on 11 August 1965, two months before McMillan stepped down as NRO Director and was replaced by Alexander Flax. This agreement returned influence back to the Defense Department. The agreement reaffirmed the NRO as a Defense Department agency, and the Secretary of Defense was responsible for its operation and had "final approval" of its budget. The DCI was no longer required to approve the Director selected by the Secretary of Defense, although the Secretary of Defense was still required to approve the DCI's selection for the Deputy Director.

This fourth agreement also established a National Reconnaissance Program Executive Committee (EXCOM), initially consisting of the Deputy Secretary of Defense as chair, the DCI, and a senior representative of the Office of the President. The EXCOM met at least twice a year, with the final meeting occurring in December, when it would approve the NRO's budget in time for it to be incorporated into the Defense Department's budget just before it was submitted to the White House. These four agreements were sometimes referred to as the NRO "charter." But, in practice, the rules that governed the NRO evolved continuously. The central issue has been whether the Secretary of Defense or the head of the Intelligence Community should have greater say over NRP planning. Some histories (especially those that deal with the NRO's early years) argue that the Defense Department prevailed over the Intelligence Community. In reality the balance has shifted back and forth. Statutes and policies created overlapping authorities.

For example, the agreements of the 1960s made the NRO a Defense Department agency and gave the Secretary of Defense authority to appoint the NRO Director. But in 1976, just a decade later, President Ford issued Executive Order 11905, designating the NRO as one of the agencies brought together into the U.S. Intelligence Community and as part of the National Foreign Intelligence Program (NFIP), which gave the DCI "full and exclusive authority" over the preparation of the NFIP budget (including the NRO). Both of these steps brought the NRO further under the authority of the DCI.

In 1986, the Goldwater-Nichols Defense Reform Act created a statutory process for reviewing defense programs—by definition, including the NRO—and an Under Secretary of Defense (Acquisition) responsible for implementing it, again strengthening the authority of the Defense Department. But the Intelligence Reform and Terrorism Prevention Act of 2004 gave the new Director of National Intelligence "milestone decision authority," or the prerogative to decide whether to allow a major intelligence program—including those run by the NRO—to proceed to the next acquisition phase, again strengthening the authority of the Intelligence Community.

The most recent turn in the relationship between the Defense Department and Intelligence Community in running the NRO occurred on 21 September 2010, when the Director of National Intelligence and Secretary of Defense signed a new memorandum of agreement. Under this agreement, which effectively serves as a new charter for the NRO, the NRO Director is responsible for managing and operating NRO programs and serves as the principal advisor to the Secretary of Defense and the DNI on overhead systems. The agreement also stipulates that the NRO Director will have direct access to both the Secretary and the DNI. The Defense Department and Intelligence Community must validate requirements for a system that uses their funds, but once a program is approved, the NRO Director has milestone decision authority for it, unless it is withheld on a "by exception" basis. (Clapper and Gates, 2010)

A certain amount of conflict between the Defense Department and Intelligence Community is an inherent part of the NRO's history, and remains a challenge for running the organization today. It will not go away because it is driven by a budget that is inevitably finite and by differences in organizational priorities, which are themselves inevitable and which also change over time.

Early NRO Programs

One of the follow-on systems to Corona was Gambit. Corona was essentially a "search" system designed to image huge tracts of land looking for unknown targets that had not yet been discovered by the U.S. intelligence community. Gambit was a high-resolution "surveillance" system designed to image known targets with much better ground resolution to discover details essential to understanding Soviet capabilities. As mutually supporting systems, Corona was designed to find new targets, and Gambit was designed to exploit new targets after they were discovered.

Even before Corona flew successfully for the first time, Land brought an Eastman Kodak proposal for a new high resolution camera to the attention of Charyk. Charyk endorsed the proposal, and it was approved for development by President Eisenhower on 25 August 1960, one week after Corona's first successful launch. Charyk assigned Brig Gen Robert E. Greer as the program's military director and tasked him to lead a program designed to produce a reconnaissance satellite capable of producing imagery with a 2-3 foot ground resolution, better than a ten-fold improvement over the Corona camera that had just been launched.

Even though Gambit borrowed much of its technology from successful Corona operations, it still took almost three years before the first launch of a Gambit system. General Greer knew that because Corona was already providing imagery and numerous other satellite projects were either being killed or not producing promised results, tolerance for failure or excessive spending on Gambit would not be acceptable. Greer moved the Gambit program along at a slower pace than Corona to both hold down costs and to ensure early success. The very first Gambit launch on 12 July 1963 was successful and produced imagery of intelligence targets, something that took 14 Corona launches to achieve.

By the time Gambit had completed 14 missions, testing and development was almost completed. In all, Gambit (KH-7) flew 38 missions through June 1967, and its follow-on, Gambit-3 (KH-8), flew 54 missions from July 1966 through April 1984, achieving ground resolutions of better than two feet. In the late 1960s, space missions had become so reliable that only three Gambit-3 missions failed to produce any intelligence, and those three failures were due to launch problems that failed to get the satellites into orbit. By the time Gambit-3 began regular launches, U.S. analysts knew, for instance, almost precisely at what rate the new Soviet T-62 tank was being delivered to Soviet

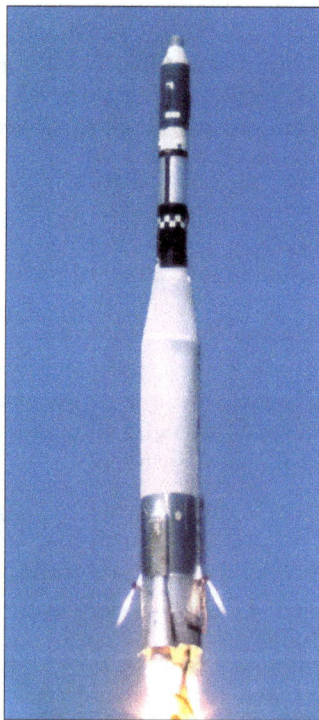

Figure 9: Gambit-1 launch

Figure 10: Hexagon Launch

tank regiments stationed along the Chinese border, and similar findings were reported for a surprisingly wide variety of aircraft, missiles, and ships. (Perry, 2012, p. 74)

Regardless of how successful Corona was, it had always been a "stop-gap" measure to be eventually replaced by a more capable system. Several different programs were investigated and subsequently cancelled in the effort to create Corona's replacement. One proposal was to replace Corona with a modified Corona and to simply update the system on a case-by-case basis as new technological developments presented improvements. This proposal was dropped after it was determined that, while cheaper than other alternatives, the capabilities of an improved Corona would never be able to meet the requirements of the nation's intelligence needs. A modified Gambit system was also proposed, but the level of technological advancements at the time made coupling the high resolution capabilities of Gambit with the broad area coverage of a Corona-type search system mutually exclusive. Another proposal, the Manned Orbiting Laboratory (MOL) program, involved a constantly-manned satellite that used astronauts to do much of the work, such as weather monitoring, target selection, and crisis adjustments, that later satellite programs would accomplish by computer. The MOL program was favored by the Air Force as a way to keep the military integrated in the space reconnaissance business, but the program was beset by schedule delays and was cancelled for fiscal reasons before the first flight test could be conducted.

After years of delays, due as much to politics and organizational tensions as to technological developments, the replacement for Corona, now known as Hexagon (KH-9), flew for the first time on 15 June 1971. By the early 1970s, launcher technology had increased as much as camera technology, so Hexagon, carried by the Titan IIID launcher, was as big as a locomotive and was a much more capable system than Corona. Hexagon contained four film-return buckets, compared to Corona's one (later upgraded to two). The first Corona satellite carried 3,000 feet of film, and all total through 145 flights, Corona returned 2.1 million feet of film; a single fully-loaded Hexagon carried over 300,000 feet of film. Hexagon was originally designed to have one satellite launched every 45 days. But due to the success of the system and the amount of imagery it provided, the limited number of photo analysts at the National Photographic Interpretation Center (NPIC) could not exploit the film in a timely manner. By the end of the Hexagon program, satellites were remaining in orbit for 6-9 months to allow analysts time to exploit the film from each bucket before the next was recovered.

The cameras aboard Hexagon acquired film with a resolution of 2-3 feet, about the same as the early Gambit cameras. But since Hexagon was a "search" system like Corona, and not a "surveillance" system like Gambit, ground resolution was never appreciably improved. Hexagon's utility was in acquiring more area, and not better resolution. Frames of Hexagon imagery covered areas as wide as 370 nautical miles, about the distance from Washington, DC to Cincinnati, OH.

In addition to the search cameras aboard Hexagon, the satellite also carried a new Mapping Camera System (MCS) on eight of its flights. The MCS collected 48,000 feet of highly accurate mapping film covering about 104 million square nautical miles. The MCS provided better than a four-fold improvement in accuracy, and more than a ten-fold improvement in resolution, over the previous best KH-5 mapping camera. This data provided far better geographic positioning and elevation information for the nation's mapping community, allowing them to produce more and better maps and targeting data for tactical and strategic weapon systems.

Hexagon flew 19 successful missions from June 1971 through October 1984. The 20th and final Hexagon mission was launched on 18 April 1986, but it experienced a booster malfunction nine seconds into flight and was destroyed, becoming the only unsuccessful Hexagon mission. Both Gambit and Hexagon were declassified by DNRO Bruce Carlson for the NRO 50th anniversary celebration on 17 September 2011.

Quill: A Radar Experiment Success

In April 1960, the U.S. Army unveiled pictures of American cities taken at night and through clouds using a synthetic aperture radar (SAR) system mounted in a small aircraft. This emerging technology was receiving significant interest from people and organizations involved in reconnaissance activities. The Air Force was particularly interested to see if this technology could be used to provide usable post-strike damage assessments without having to wait for appropriate conditions for optical sensors.

In late 1962, Charyk designated Maj David D. Bradburn (who would later become a Major General and head NRO's Program A) to lead a project named Quill to determine if collection of usable SAR imagery from satellites was feasible. Because the program was purely experimental to test the feasibility of technology and not an operational program to build a series of satellite collection platforms, Bradburn curtailed the objectives of the program – a significant departure from most military programs of the day that usually expanded well beyond their initial plans.

Figure 11: Quill Vehicle

Using "off-the-shelf" equipment and technology, along with experienced contractors Goodyear Aerospace and Lockheed Missiles and Space Company, Bradburn was able to quickly and efficiently get the program off the ground. Quill collected radar returns on tape spooled in the satellite, as well as transmitting the data back to collection sites on earth. The first (and only) Quill launch occurred on 21 December 1964. The satellite worked so well that a second planned launch was cancelled, since all of the program's objectives had been met during the first launch.

In the final evaluation of the experiment, it was found that usable SAR imagery could indeed be collected from satellites. However, the resolution of the Quill imagery was relatively poor, and it would be many years before the IC would be able to build a usable radar satellite; it was not until 15 October 2007 that the DNI declassified the fact that the U.S. operated an effective radar satellite reconnaissance program. Due to the limited scope of the experiment and Maj Bradburn's leadership, the Quill program was the only early NRO program to be completed on time and under budget. The DNI approved the declassification of the Quill program in November 2009.

Verifying Arms Control

Arms control—specifically, ceilings on strategic weapons—became the lynchpin of U.S.–Soviet relations after President Nixon took office in 1969. The feasibility of a treaty depended on whether U.S. intelligence could detect potential Soviet violations. Several events had combined to put satellite reconnaissance at the center of the negotiations.

First, the Soviets had themselves established the right to orbit satellites across national boundaries by launching Sputnik 1 in 1957. Then, during the early 1960s the Soviets gained parity with U.S. reconnaissance capabilities when they developed their own intelligence satellites. As Soviet Premier Nikita Khrushchev reportedly told Charles DeGaulle at the 1960 Paris Four Power Summit, "Airplanes nyet; Sputniks, OK."[4] By 1963 Khrushchev told an interviewer that on-site inspection was no longer a barrier to an arms control agreement because the "function can now be assumed by satellites." (Klass, 1972, p. 31)

Because both the United States and the Soviet Union accepted satellite reconnaissance (or realized they could not prevent it), overhead systems became the natural solution for monitoring an arms control agreement that both governments wanted. Thus, Article V of the 1971 interim agreement from the Strategic Arms Limitation Agreement (SALT I) said that "each Party shall use national technical means of verification," referring to IMINT and SIGINT, usually collected by overhead systems.

---

4   See Richard Garwin's account in McDonald, R.A., Ed. (2002). "Recollections of the Pioneers and Founders of National Reconnaissance." Bethesda, MD. *American Society for Photogrammetry and Remote Sensing*, p. 20.

Arms control became a critical factor in the planning of NRO programs during this period, as John L. McLucas became NRO Director in 1969. The various treaties established specific numerical ceilings or design limits on each country's weapons. These drove requirements for NRO systems. For example, SALT had different limits for large ICBMs and small ICBMs, and stipulated that neither country could enlarge its missile silos by more than 15 percent. NRO systems thus had the task of distinguishing between large and small missiles, and assessing the volume of Soviet silos. The treaty language, in effect, became requirements for the capabilities of NRO satellites.

Beginning the Reduction of Secrecy

Arms control also triggered a process in which the NRO and its activities gradually became less secret. When the Nixon administration prepared SALT for Senate ratification in 1971, senior officials debated whether to acknowledge that satellite reconnaissance was a "national technical means" referred to in the treaty. President Nixon decided not to, concerned that countries other than the Soviet Union might object to U.S. surveillance. The Senate ratified the treaty without explicitly discussing the exact definition of national technical means, or the NRO and its capabilities.

As arms control became more contentious in the 1970s, however, there was more disagreement over whether the United States could monitor an agreement effectively. This was the main reason President Carter officially acknowledged U.S. satellite imagery systems on 1 October 1978. Near the end of a seemingly routine speech about U.S. achievements in space at Kennedy Space Center, President Carter mentioned that "photoreconnaissance satellites have become an important stabilizing factor in world affairs in the monitoring of arms control agreements," and that the United States would continue to develop them. The Senate was considering SALT II for ratification, and the comment was intended to persuade skeptics and give officials greater leeway in discussing U.S. intelligence capabilities.

Additional information about the NRO surfaced during this period at the espionage trial of William Kampiles in November 1978. Kampiles, a former CIA employee, was convicted for stealing a manual to the KH-11 system and selling part of it to Russian officials in Athens. The government acknowledged during the trial that the KH-11 was a satellite-based electro-optical imagery system. Eight years later, an actual example of overhead imagery appeared in the press when Samuel Loring Morison, an analyst at the Naval Intelligence Support Center, sold a KH-11 image he had stolen to a defense publication. (Morison was also convicted.)

The NRO had pioneered electro-optical systems in the 1970s when it became clear that both political officials and military planners needed a more responsive photoreconnaissance satellite. The Intelligence Community had not been able to retrieve satellite imagery of Soviet forces preparing to invade Czechoslovakia in August 1969 until after the crisis was over. Similarly, the Intelligence Community was surprised by the strike on Israel by Egypt and Syria in October 1973 in part because the attackers acted faster than the NRO's imaging systems could respond.

The NRO greatly expanded its capabilities during this period. Electro-optical systems introduced when James Plummer served as NRO Director and then fully implemented under his successor, Thomas Reed, were a major breakthrough that revolutionized satellite photoreconnaissance, totally replacing film return systems. The NRO also developed new SIGINT collection systems and programs to support military operations under Robert Hermann, who served previously at the National Security Agency.

Launch Vehicle Problems

In the mid-1980s, a string of events demonstrated how much the nation had come to depend on NRO systems, despite the high risk associated with developing, launching, and operating the nation's reconnaissance satellites.

During the 1970s U.S. policy made the Space Transportation System—better known as the Space Shuttle—the primary means for launching government payloads. The Shuttle promised lower costs and more frequent launch opportunities. NRO Director Hans Mark, an advocate of reusable launch vehicles, endorsed this policy. The NRO (along with other government agencies) began to wind down their expendable launch vehicle programs, and began to optimize their satellites for launch on the Shuttle.

Then, on 28 August 1985, a Titan 34-D launched from Vandenberg Air Force Base failed during ascent, resulting in the loss of the payload. Five months later, on 28 January 1986, the Space Shuttle *Challenger* broke apart catastrophically during launch. On 18 April 1986, another Titan 34-D failed when a solid fuel booster exploded shortly after liftoff at Vandenberg.

The August 1985 launch failure grounded the Titan vehicle for eight months while investigators analyzed the cause. The April 1986 failure damaged the launch pad, taking it out of operation for more than a year. After the *Challenger* loss, the Shuttle did not return to operation until September 1988. Although there was no immediate threat to the NRO's mission capability, these launch failures highlighted the potential fragility of its constellation, and thus the need for reliable, redundant launch systems. The NRO had come to rely on smaller numbers of satellites, partly because satellites had become bigger and more reliable, and thus had longer operational lives. The new satellites were more capable, but because they were fewer in number, the loss of one or two represented a major loss in total capacity for the NRO. Since most satellites at the time had to be launched on a specific launch vehicle or from a specific pad, the grounding of a vehicle or the loss of a pad could cause significant complications to NRO operations.

The Shuttle presented its own issues. Following the investigations of the *Challenger* loss, the Reagan administration decided that it was unwise to risk the Shuttle and the lives of astronauts to launch satellites, and the NRO returned to relying mainly on expendable launch vehicles. Fortunately, when the Shuttle program began to encounter delays in the early 1980s, NRO Director Edward "Pete" Aldridge had proposed retaining expendable vehicles as an alternative to the Shuttle; it was in large part because of Aldridge's efforts that the Titan production line continued to operate.

The launch crisis of the late 1980s was one reason for the decision to develop the Evolved Expendable Launch Vehicle (EELV). The goal was to create a more reliable, flexible, and less expensive family of launch systems by building on the experience of the existing Atlas, Titan, and Delta vehicles. The NRO began announcing launches on an unclassified basis beginning in 1996. The Air Force awarded contracts for the EELV to Lockheed Martin and Boeing in late 1998, which developed the Atlas V and Delta IV launch vehicles, respectively. Even with the new generation of vehicles, launch capability remains a critical link, and the NRO is often stressed to maintain its planned constellation because of the limited capacity of the ground infrastructure. (Carlson, 2010)

Desert Storm, Post-Cold War Drawdown, Declassification, and Controversy

Operation Desert Storm, the 1991 campaign to liberate Kuwait after its invasion by Iraq, marked the first "high tech" war. NRO imagery and signals intelligence systems played an important role in the decisive U.S.-led victory. But the war also showed that NRO systems, which had been optimized to monitor arms control agreements, were often ill-suited to support combat operations. Gen H. Norman Schwarzkopf, commander of the Coalition forces, told members of Congress that damage assessments were "one of the major areas of confusion." (Moore, 1991) He said that CIA and DIA analysts, using satellite imagery, underestimated bomb damage by Coalition strikes, causing him to err too often on the side of caution as the conflict evolved.

Schwarzkopf also said that intelligence arrived too slowly. "The intelligence community," he said, "should be asked to come up with a system that will, in fact be capable of delivering a real-time product to a theater commander when he requests that." The House Armed Services Committee later cited "significant problems in intelligence support" in which imagery "was often late, unsatisfactory, or unusable." It predicted that "The need for intelligence will grow as next generation weapons enter the inventory. And as the sophistication of weapons increases, deficiencies in intelligence support will proportionally constrain their effectiveness." (House Armed Services Committee, 1991)

Reflecting the experience of the Gulf War, President Clinton issued Presidential Decision Directive 35 in March 1995. This directive codified support to military forces as a new top intelligence priority. Throughout the decade, the need to locate specific targets and provide data directly to U.S. warfighters shaped the planning, design, and operation of NRO systems.

Yet even as demands for NRO support grew, budgets got tighter. During the 1980s and the early 1990s, the NRP had grown annually, even accounting for inflation. With the Cold War over, Democrats and Republicans both supported a "peace dividend" by cutting defense and intelligence spending. Between Fiscal Year 1990 and 1997 the budget of the National Foreign Intelligence Program declined by 14 percent. (see Combest, 1996)

This pressed the NRO to reduce costs by consolidating programs, building smaller, less expensive satellites, and adopting contracting practices that reduced paperwork and gave contractors incentives to work more efficiently. All of these choices, together with the need to support a growing number of users and missions, led to decisions—and controversies—that affect almost every aspect of NRO operations even today, two decades later.

Consolidating the "Alphabet Programs" and Declassifying "Fact Of"

In 1991 DCI Robert Gates appointed a commission chaired by former Lockheed CEO Robert Fuhrman to review the structure and operation of the NRO. Many of the Fuhrman Commission's recommendations had circulated in the satellite reconnaissance community for some time. Two years earlier NRO Director Martin Faga had commissioned a study panel headed by RADM Robert Geiger and Barry Kelly, former Special Assistant to the President, to consider how to improve efficiency at the NRO. The end of the Cold War provided the opportunity to act on these ideas.

One step was to consolidate the "Alphabet Programs" into a structure with less redundancy and more central control. NRO Director Faga accepted the recommendations, and they were adopted as part of National Security Directive 67, which President George H.W. Bush signed on 30 March 1992. The reorganization took effect on 31 December 1992. Programs A, B, and C were replaced with three functional directorates, all to be located at the NRO's new headquarters complex to be constructed in Chantilly, Virginia. The three directorates were:

- The Signals Intelligence Systems Acquisition and Operations Directorate (SIGINT), responsible for acquiring and operating satellites that collect communications, telemetry, and other electronic emissions.

- The Imagery Intelligence Systems Acquisition and Operations Directorate (IMINT) responsible for acquiring and operating satellites that collect electro-optical imagery.

- The Communications Directorate (COMM), responsible for the NRO's information technology and communications systems, as well as security for both space-based and ground-based communications used by military forces, the Intelligence Community, and other government users.

The NRO established the Advanced Systems and Technology Directorate (AS&T) as a fourth directorate in March 1997, responsible for new satellite reconnaissance research and development.

The Fuhrman Commission had also recommended declassifying the "fact of" the NRO's existence. As we have seen, by 1978 satellite reconnaissance was already recognized as a "national technical means" to monitor arms control, and by 1992 more countries were developing satellite reconnaissance systems. Companies such as France's SPOT even sold imagery products commercially that were comparable in resolution to Corona imagery from the 1960s.[5]

Also, the emphasis on support to military operations that followed Desert Storm meant more people would have access to satellite reconnaissance products. It would be hard to plan and disseminate these products without at least acknowledging the fact of an organization that produced them. Finally, the Senate Select Committee on Intelligence (SSCI) informed the Intelligence Community that it wanted to declassify the existence of the NRO unless the President could certify that it would cause "serious damage" to the nation (the standard for classifying information as "Secret" under Executive Orders). The existence of the NRO was declassified on 17 September 1992.

The Funding and the Headquarters Controversies

After the Fuhrman Commission had recommended consolidating the newly created directorates into a single headquarters, numerous NRO documents throughout the early 1990s referred to the coming "collocation" in the Washington area. Some referred to the site of the new headquarters, planned for the Westfields office complex in Chantilly, Virginia, although this location remained classified.

On 8 August 1994, SSCI Chairman Dennis DeConcini (D-Ariz.) and Vice Chairman John Warner (R-Va.) released a letter saying that they were "shocked and dismayed to learn that the cost of the new NRO headquarters at Westfields may reach $350 million by completion, nearly double the amount most recently briefed to the committee." They said that "the total anticipated cost was never effectively disclosed to our committee…" Sen. Warner said "I was absolutely astonished at the magnitude and proportions of this structure."

Members of the House Permanent Select Committee on Intelligence (HPSCI) disagreed with their Senate counterparts. HPSCI Chairman Larry Combest (R-Tex.) said "Charges of CIA or NRO deception are absolutely erroneous." Ranking minority member Norman Dicks (D-Wash.) said that his staff had been informed about the cost of the building and that "It's an open and shut case that they were properly briefed." DCI R. James Woolsey insisted that documents submitted to the SSCI over four years had shown the size and cost of the new facility.

---

5   The U.S. government began to purchase intelligence imagery from commercial operators beginning in 2000, following the launch of Ikonos by Space Imaging. Since then the National Geospatial-Intelligence Agency has greatly expanded this initiative, and the U.S. Government has licensed commercial operators to operate even more capable systems.

NRO Deputy Director, Jimmie Hill, testified to the SSCI that the NRO had treated the building as part of the costs of supporting the operation of the agency, rather than as a separate project, consistent with an understanding he had reached earlier with staffers of the committee. Even so, NRO Director Jeffrey Harris told the committee members two days later that, in hindsight, "the building costs should have been broken out specifically" in the budget (Thomas, 1994; Weiner, 1994a; Weiner 1994b; Pincus, 1994; and Laurie 2001).

A joint review carried out by the Defense Department and CIA found no intent by NRO officials to mislead Congress and that the NRO had provided cost data to Congress when specifically requested. It also found that Congress approved the purchase of property for a new headquarters and funds for starting construction when it approved the reorganization of the NRO, and that the new building was no more expensive than one would expect for a structure its size. However, the review also concluded that the NRO failed to follow Intelligence Community budgeting guidelines, and that by leaving funds for the building in the agency's "base" budget for operations, the total cost of the project was unclear to an outside observer. Whatever the merits of the case, the incident fixed an unfavorable image of the NRO in the public's mind. This was compounded the following year when another spending controversy erupted.

The SSCI had begun to investigate the NRO's accounting practices in 1992 when it determined that the agency had accumulated unusually large sums in carryover accounts—money appropriated for a program but which the NRO had not yet spent and held for the following year. The NRO had assured the SSCI that it would eliminate such excessive "forward funding." In September 1995 Senator Arlen Specter (R-Pa.), now chairman of the SSCI, said that the NRO had failed to do so (Specter, 1995). An initial survey that summer estimated that carryover funds across the entire NRO totaled $1.7 billion. After further review, this estimate rose to $3.7 billion. (Laurie, 2001)

These changing estimates and the NRO's inability to provide a single, firm estimate of the total carryover funds was as damaging as their initial discovery. Some journalists reported that the NRO had "lost" the money in "a complete collapse

Figure 12: NRO Headquarters Under Construction

25

of accountability." Another SSCI member, Sen. Richard Bryan (D-Nev.) said that there was "rampant mismanagement" at the NRO. DCI John Deutch said that he was unaware of the surplus, as did White House Chief of Staff Leon Panetta. (Weiner, 1996; Bryan, 1995)

However, the SSCI's Vice Chairman, Robert Kerrey (D-Neb.) said that, based on the hearings that the SSCI had held, the public accounts of the forward funding were not accurate. Kerrey said that the funds were maintained in accord with DOD regulations, and that Defense Department officials had known about the funds since at least 1989, when the DOD Inspector General audited the NRO and agreed with the size of the fund and its method of accounting. (Laurie, 2001)

The accumulated funds were a result of several factors. As satellite technology matured, NRO satellites grew larger, became more complex and took longer to build. As noted previously, they also became more reliable and thus often lasted longer, although it was hard to forecast how long. By the 1990s, planning, building, and launching a satellite had become a process that extended over several years, with more uncertainty in knowing exactly when the satellite would be needed on orbit.

If it appeared that a satellite was not needed when originally planned, the NRO would hold off procuring the replacement. It put most of the funds in reserve and procured just enough hardware to ensure that it would not run short of critical components. It also funded enough activity to sustain its "industrial base," which might atrophy if programs were simply deferred and skilled contractor personnel had to find other work. There was an implicit assumption that the laws of probability would rule, so that as some launches failed and some satellites expired earlier than planned, the funding flow would balance. The "forward funding" provided a buffer. In any case, the NRO did not spend the funds on programs other than what had been authorized.

One reason the NRO was unable to quickly provide Congress a specific figure for exactly how much spending authority it had accumulated was because of the organization's history. The NRO had inherited an assortment of highly compartmented budgeting systems from Programs A, B, and C—each based in a different department or agency—the Air Force, CIA, and Navy. Each had its own procedures for managing secret operations. As a result, there was no single accounting system for NRO officials to readily monitor unspent funds across programs. Indeed, few persons had the authority to gather all of the information.

In addition to all of this, some officials and legislators seemed to expect NRO programs to operate like a typical Defense Department weapons acquisition program. In fact, there are important factors that make the two very different.

A typical military acquisition program has a development phase and a production phase. The development phase might involve building one or two prototypes, and inevitably requires solving unexpected problems as they are encountered. Budgets for such development efforts usually contain a margin to accommodate the uncertainty that comes with building a new system. Later, in the production phase, budgets require less margin as the contractor gains experience and costs become more predictable.

NRO programs usually do not follow this pattern. The NRO has historically built satellites in small numbers; the main exceptions were the early film return IMINT satellites. So, the first satellite of an NRO program was in effect, both the prototype and a production item. It is also common for the NRO to make changes—often major—from one satellite in a series to the next to add a capability or fix a deficiency. In effect, NRO programs never really went into production, at least in the way most military systems did; they were always in the development phase. This was why NRO program managers had typically added 20 to 30 percent to a contractor's bid—a level of margin acceptable for a development program, but far more than one would expect for a typical DOD acquisition. (See Fitzgerald, 2005a, 2005b, 2005c; Kohler, 2005; and Nowinski and Kohler, 2006)

As a result of the forward funding controversy, Secretary of Defense William Perry and Director of Central Intelligence John Deutch directed Director Harris and Deputy Director Hill to resign. Congress passed legislation to recoup the carry-over funds, which it applied to U.S. operations in the Balkans, the B-2 bomber program, and other defense projects. (Morgan and Pincus, 1996)

The NRO began to develop a corporate accounting system to avoid similar situations in the future. Perry appointed a new NRO Director, Keith Hall, who ordered a comprehensive review of NRO finances (Hall had previously served at the SSCI, as well as the Office of Management and Budget, the Office of the Secretary of Defense, and the CIA). Hall led an effort to set up a new financial system so that today NRO acquisition operates more like other DOD organizations. The current system requires programs to use independent cost estimates when formulating budget requests. It also uses metrics to measure progress in a program, and then manages budget margins to limit carryover funds to a few months' worth.

In many respects, the practices criticized in the forward funding controversy—a highly compartmented process that minimized administration in favor of success on a tight schedule—were the very practices that had originally justified the NRO's establishment. President Eisenhower had taken the WS-117L out of the Air Force chain of command, and then assigned what became Corona to the CIA because he wanted a classified, more urgent program that was not bound by the usual constraints of the federal acquisition process. Several administrations and Congress had supported this approach because in the 1950s and 1960s they believed assessing the Soviet strategic threat was so important. In the 1970s and 1980s monitoring arms was considered just as important. But by 1995, events made it clear that support for this approach had waned.

Redesign of the Overhead Architecture

In the mid-1990s the NRO began an effort to fundamentally redesign its overhead architecture. Deputy Director Jimmie Hill described this effort in 1994. "At no other time since the creation of the NRO," Hill told legislators, "has the government embarked upon such a significant change in all of its satellite capabilities." (Hill, 2001)

Several factors were behind the redesign. First, and possibly most important, the general drawdown in defense and intelligence spending pressed the NRO to reduce costs. In 1996 a panel commissioned by DCI John Deutch at the direction of the HPSCI concluded that, in general, it was possible and desirable to build smaller, less expensive reconnaissance satellites (Deutch, 27 June 1996). Also, some critics claimed the NRO was losing its reputation for innovation, and pressed it to introduce systems as ground-breaking as Corona and Grab had been in the 1960s, and electro-optical systems had been in the 1970s.

The SIGINT component of NRO's plan, the Integrated Overhead Signals Intelligence Architecture (IOSA), proceeded relatively smoothly. IOSA was mainly a process of consolidating payloads that had been operated in similar orbits onto a smaller number of satellites and integrating the ground stations. The satellites themselves were either incrementally improved versions of existing systems, or new systems using contractors that had many years of experience building the earlier systems. The relay satellite component of the plan also proceeded smoothly, as it was also an incremental change.

The IMINT components—the Future Imagery Architecture (FIA)—were much more problematic. FIA was the first IMINT system NRO developed after the dissolution of the alphabet programs. The reorganization had broken up the Program A and Program B teams that had developed the earlier systems. The NRO required the FIA contractor to build the system to a specified, not-to-exceed cost (a direct result of policies resulting from the forward funding controversy), gave more responsibility to the contractor to manage the program (to improve efficiency and

Figure 13: NRO Headquarters Today

reduce costs), and encouraged the contractors to propose smaller, lighter vehicles (again, to reduce costs).

The NRO announced on 3 September 1999 that Boeing had won the FIA contract. Hall, NRO Director at the time of the award, recalled later that Boeing proposed a more innovative design than its competitor, Lockheed Martin. He also recalled that the NRO doubted Lockheed Martin could meet the cost ceiling with the system it proposed, but believed Boeing could (Taubman, 2007). The program soon encountered technical problems, leading to a major spacecraft redesign, resulting in delays and overruns. The program lacked the margin the NRO had used in the past to accommodate such problems, and, to make matters even worse, the NRO had difficulty assessing the problems and directing corrective action because so much responsibility for the program had been given to the contractor.

FIA became a continuing problem for the NRO. In May 2003, a joint Defense Science Board and Air Force Scientific Advisory Board task force found that FIA was "significantly underfunded and technically flawed." In 2005, DNI John Negroponte decided the issue and terminated FIA as it had originally been constituted. The NRO began developing a new strategy for IMINT. In 2009 NRO Director Bruce Carlson described in general terms a new electro-optical system, Next Generation Electo-Optical (NGEO) that will be a lower-risk modular system, capable of being modified in increments over its lifetime. (Carlson, 2009)

As the 1990s were ending, Congress had chartered a commission to study the future of the NRO. The commission, chaired by Senator Robert Kerrey (D-Neb.) and Representative Porter Goss (R-Fl.), reported in November 2000 that the NRO was facing the challenge of supporting a growing number of users and missions with a budget that had not grown proportionately. The commission concluded that the NRO needed more funding to carry out its mission; a more effective means to prioritize its limited budget; and a new capability to do high priority, tight security projects, as it had in the 1960s. Those recommendations may have given the NRO the flexibility to direct the FIA program into a more successful outcome.

The NRO in the Twenty-First Century

Fifty years after its establishment, the NRO has become a global organization, managing a complex system of satellites and ground stations that provides intelligence support to an ever-expanding number and variety of users. The NRO today operates a highly integrated architecture of satellites for signals intelligence, imagery intelligence, and communications, in addition to its network of ground stations.

NRO signals intelligence satellites, tracing their origin to Grab, continue to collect a variety of forms of information across the electromagnetic spectrum. These include FISINT ("foreign instrumentation signals intelligence" or data collected during the test or operation of aircraft, missiles, or other systems); COMINT (communications intelligence taken from voice, text, or pictoral transmissions); and ELINT (electronic intelligence from non-literal transmissions, such as radar).

NRO imagery systems, tracing their origin to Corona, today include both electro-optical and synthetic aperture radar satellites. These satellites provide a near-real-time capability and provide U.S. military forces information for indications and warning, as well as for planning and conducting military operations. In addition, these imaging systems can be used to collect scientific and environmental data and data on natural or man-made disasters.

The NRO constellation today also provides a capability to collect various forms of MASINT, or "measurement and signature intelligence." MASINT is based on the analysis of characteristics associated with specific targets or classes of targets. MASINT has generally become more important to U.S. military forces because it is used to characterize foreign weapons systems, and is also used in conjunction with IMINT to program precision-guided munitions, which have become a larger part of U.S. military operations. MASINT collected from satellites also provides intelligence used for indications and warning.

One challenge that the NRO grapples with today is the increasing age of its satellite systems. Currently some NRO satellites are more than 20 years old. This is partly good news because it reflects the improvements in satellite lifetime that the NRO has achieved. The design margins originally needed to meet minimum requirements for reliability have typically allowed a satellite to greatly exceed its planned lifespan. However, it also means that the level of service that the NRO currently provides depends on an aging, and thus increasingly fragile, constellation. (Carlson, 2010)

To support this constellation, the NRO depends on a network of ground stations. This network includes the Aerospace Data Facility–East at Ft. Belvoir, Virginia; the Aerospace Data Facility–Southwest at the White Sands Missile Test Range, New Mexico; and the Aerospace Data Facility–Colorado at Buckley Air Force Base, Colorado. Each is a multi-mission facility that supports worldwide defense operations and the collection, analysis, reporting, and dissemination of intelligence information for multiple agencies.

The NRO also maintains a presence at several locations overseas. These include the Joint Defense Facility Pine Gap in Alice Springs, Australia and RAF Menwith Hill, in Harrogate, United Kingdom. The NRO supports joint missions at these locations through the provision of technical systems and shared research and development. The NRO's participation is achieved with the consent of the host governments and contributes to the national security of the countries involved.

In addition to its intelligence collection systems, the NRO maintains an extensive global communications network that supports both NRO operators and other military and intelligence users. The NRO's communications infrastructure includes, for example, its encrypted satellite data relay system and messaging systems essential for the organization and its partners to function, such as the Special Operations Communications (SOCOMM) system.

Support to Current Military Operations

The NRO has played a key role in operations against al-Qa'ida and other terrorist organizations, as well as U.S.-led military operations against insurgencies in Iraq and Afghanistan. These new adversaries often operate as dispersed, clandestine networks and use geography to their advantage by blending in among the local population, hiding in isolated, rugged locales like the Afghan-Pakistani border, or in hostile, ungoverned regions of Somalia or Yemen. Also, these adversaries are successful in their use of technology; the Internet is a key recruitment tool, and a favored weapon is the improvised explosive device (IED).

In these new conflicts, U.S. forces often must find specific individuals—terrorist leaders, financiers, bomb-makers and other "high value targets" (HVTs)—or specific objects, such as WMD components. Often the NRO has had the only collection capability that could provide the intelligence that U.S. officials and military forces require. To do this, the NRO has had to rethink its operations to deal with these new threats.

The NRO routinely collects intelligence for U.S. military operations, and in today's environment this support is more likely to be "multi-INT," combining overhead intelligence with other data. The NRO's workforce today thus includes personnel from throughout the Intelligence Community. NSA and NGA personnel are often assigned to NRO facilities. Similarly, because U.S. forces are most likely required to operate as members of a coalition, the NRO workforce also includes representatives from the United Kingdom, Australia, Canada, and New Zealand. NRO personnel themselves today more frequently deploy "downrange" with more than 40 people typically deployed in combat theaters. (Sapp, 2010)

Figure 14: IED Munitions in Baghdad

In addition to intelligence collection, the NRO also provides communications support to U.S. and Coalition forces. For example, in one recent initiative NSA, NGA, and the NRO combined their capabilities to develop an integrated counter-IED capability. One of the NRO's key contributions in this effort was providing the communications backbone for the system.

Because today's threats can change tactics and methods so quickly, the NRO has put greater emphasis on the ground segment of its systems. Though designing and building a new satellite today can require several years, it is often possible to develop a new data processing system or software tool in a few months to exploit data from the existing satellite constellation. By focusing on the ground segment of a system, NRO can make more frequent modifications and add additional capabilities more easily. This was one reason why NRO Director Scott Large established a new Ground Enterprise Directorate (GED) in 2008, incorporating the ground system elements of the SIGINT and IMINT Directorates, along with other NRO components. GED works closely with NGA and NSA, as well as NRO's own Advanced Systems and Technology (AS&T) Directorate and other NRO components to develop new multi-INT processing applications. GED also works with the COMM Directorate to find new ways to use the NRO communications infrastructure, and the new Mission Support Directorate (MSD) to better understand the specific needs of users.

Support to Arms Control and Other Missions

In addition to supporting military operations such as those in Iraq and Afghanistan, the NRO continues to monitor arms control treaties and other international agreements. Although the first Strategic Arms Reduction Treaty (START-1), which went into effect in 1994, began a process in which on-site inspections have played a larger role in verifying compliance, overhead systems are still important to provide U.S. officials assurance that treaty partners are not concealing undeclared facilities or capabilities.

Today the NRO also supports many domestic users. NRO imagery has been available to users outside the national security community since October 1975, when President Ford chartered the Civil Applications Committee (CAC). Through the CAC, the NRO can provide products to the Departments of Agriculture, Commerce, Interior, Transportation, Health and Human Services, NASA, the Federal Emergency Management Agency, the Environmental Protection Agency, the National Science Foundation, and the Coast Guard. The CAC process ensures that NRO products can be used more widely, while also ensuring compliance with U.S. privacy statutes and regulations. During the 1990s NRO imagery was also used in the MEDEA project, an Intelligence Community pilot study to assess how intelligence projects could be used in long-term environmental studies.

More recently, the NRO has provided intelligence and information to law enforcement, counterterrorism, and border protection organizations. The NRO has also supported security planners for major public events to ensure public safety. U.S. reconnaissance satellites have provided critical information to first responders and relief operations during natural disasters. After Hurricane Katrina struck the southeastern United States in August 2005, for example, the Federal Emergency Management Agency and the Army Corps of Engineers used NRO imagery to assess flooded areas and identify the location of hazards.

New Requirements for Information Sharing, New Approaches to Secrecy

As the users of NRO-derived intelligence grew, it became impractical for the NRO to conceal the basic features of its systems and operations as it had in its early years. However, some experts believed that opening up the NRO had the unintended effect of making it harder to protect truly sensitive capabilities, and this, in turn, made it harder for the NRO to develop the kinds of breakthrough systems that it was known for in its early years.

The NRO Commission captured this sentiment in its November 2000 report, recommending that the NRO strike a more nuanced balance between openness and secrecy. NRO Director Peter Teets and his successor, Donald Kerr, adopted these recommendations with an approach that would ultimately make most NRO program information more widely available, while compartmenting a much more limited set of information related to new or especially sensitive systems.

Under its new policy, the NRO put most information that had been protected in the BYEMAN control system into the TALENT KEYHOLE control system, a larger compartment that most NRO consumers with clearances could already access. The new policy then established a new, smaller, and more limited control system, RESERVE. This approach has allowed the NRO to develop new systems and technology with greater security and speed, and then make information about each system available to larger groups of users gradually, as each becomes more widely used and more individuals are aware of its existence.

Looking to the Future

When the NRO was established in 1961, several factors combined to create a unique organization. Key technologies—launch vehicles and sensors—had just reached a level of maturity that allowed some visionaries to imagine how they might solve the problem of monitoring the Soviet Union. Because the Soviet nuclear threat was so menacing, U.S. officials were prepared to give the NRO enormous discretion in finance and management. Since relatively few individuals needed access to satellite intelligence and because the technology was so sensitive, these officials were also prepared to allow the NRO to work in strict secrecy.

Today, more than 50 years later, the nature of the threat has changed. Instead of a single, slowly changing, existential threat, now the United States faces a variety of changing threats. Many can cause significant harm, but nothing on the scale of a Soviet nuclear strike. American politics has changed, too. Congress expects much more information, and more influence over national security decisions than was the case in 1961. It is unlikely that any government program could enjoy the kind of autonomy and maintain the blanket secrecy that the NRO did in its early years.

Moreover, after five decades the NRO supports many more users. Overhead reconnaissance is one of several capabilities that must be integrated into intelligence. In addition to supporting national security, the NRO also fulfills requirements to monitor the natural environment. National reconnaissance systems can monitor desertification issues, measure crop sizes, warn against volcanic eruptions, and track geological and glacial change. The NRO's capabilities helped assess the damage of the 2004 tsunami in Southeast Asia and Hurricane Katrina in 2005. They helped save lives by supporting responders fighting wildfires in the American West in 2007 and 2008, and continue to do so today.

Many of the technologies the NRO pioneered are now commercially available and widely known. The NRO's innovations have impacted our daily lives by improving the technology many of us use regularly. NRO research has contributed to the development of high definition television, GPS systems, digital imaging, video recording, and cellular phones. It has impacted the commercial camera and film industries. NRO technologies are also used to fight breast cancer by improving both its detection and treatment.

As a result, the NRO is constantly challenged to identify when and how it can make its unique contribution to U.S. security with new capabilities that our adversaries are unaware of, or at least unable to counter. Partly because the NRO has been so successful, more users depend on it. This has raised the potential danger of a failure. This makes it harder for the NRO to take some of the risks it took in its early years. Yet few organizations are better situated for taking those risks. The NRO has its own authorities, budget, access to personnel, and mission.

After fifty years, the challenge for the NRO is to maintain the reliability and contain the costs of its current systems, while at the same time providing the opportunity and challenge that attracts the nation's top minds to imagine new ways to protect American security with overhead reconnaissance.

References

Bryan, R. (29 September 1995). *Congressional Record.* S14787.

Burnett, M. G. (April 2012). *"Hexagon (KH-9): Mapping Camera Program and Evolution."* Chantilly, VA: NRO, Center for the Study of National Reconnaissance.

Carlson, B. (21 October 2009). Comments at GEOINT Symposium, San Antonio, Texas.

Carlson, B. (14 April 2010). Comments to National Space Symposium. Colorado Springs, Colorado.

Chester, R. J. (April 2012). *"A History of the Hexagon Program."* Chantilly, VA: NRO, Center for the Study of National Reconnaissance.

Clarke, A.C. (February 1945). "V-2 for Ionosphere Research?" *Wireless World.*

Clausen, I., Miller, E. A., et al. (April 2012). *"Intelligence Revolution 1960: Retrieving the Corona Imagery that Helped Win the Cold War."* Chantilly, VA: NRO, Center for the Study of National Reconnaissance.

Clapper, J.R., and Gates, R.M. (21 September 2010). Memorandum of Agreement Between the Secretary of Defense and the Director of National Intelligence Concerning the National Reconnaissance Office.

Combest, L. (22 May 1996). Floor speech presenting the Intelligence Authorization Act of Fiscal Year 1997. *Congressional Record.* H5391.

Committee on Armed Services. U.S. House of Representatives (30 March 1992). *Interim Report.*

Defense Science Board/Air Force Scientific Advisory Board (May 2003) Report of the Joint Task Force on Acquisition of National Security Space Programs. Washington, D.C.: Office of the Under Secretary of Defense for Acquisition, Technology, and Logistics.

Deutch, J. (27 June 1996). Letter to Rep. L. Combest, Chairman of the House Permanent Select Committee on Intelligence.

Douglas Aircraft Company, Inc., Santa Monica Plant Engineering Division. (2 May 1946). *Preliminary Design of an Experimental World-Circling Spaceship.* Report N. SM-11827, Contract W33-038.

Fitzgerald, D.D. (2005a). "Commentary on Kohler's 'Recapturing What Made the NRO Great' Updated Observations on 'The Decline of the NRO.'" *National Reconnaissance.* 2005-U1. 59-65.

Fitzgerald, D. D. (2005b). "Commentary on 'The Decline of the National Reconnaissance Office' The NRO Leadership Replies." *National Reconnaissance.* 2005-U1. 45-49.

Fitzgerald, D. D. (2005c). "Risk Management and National Reconnaissance From the Cold War Up to the Global War on Terrorism." *National Reconnaissance.* 2005-U1. 9-18.

Greer, K.E. (Spring 1973). "Corona," *Studies in Intelligence,* Special Supplement. 1-37.

Haines, G.K. (1996). *The National Reconnaissance Office: Its Origins, Creation, and Early Years.* Washington, D.C.: National Reconnaissance Office.

Hall, R.C. (1997). "Post War Strategic Reconnaissance and the Genesis of Project Corona," in McDonald, R.D. ed. *Corona: Between the Sun and the Earth.* Bethesda, MD: American Society for Photogrammetry and Remote Sensing.

Hall, R.C. (1999). "The National Reconnaissance Office: A Brief History of its Creation and Evolution." *Space Times.* (March-April 1999).

Hall, R.C. and Laurie, C.D., Eds. (2003). *Early Cold War Overflights: Symposium Proceedings, Vol. 1: Memoirs.* Washington, D.C.: Office of the Historian, National Reconnaissance Office.

Hill, J.D. (26 April 1994). "NRO Presentation to Defense Subcommittee, Senate Appropriations Committee, quoted in Laurie, C. (2001) *Congress and the National Reconnaissance Office.* Chantilly, VA: National Reconnaissance Office History Staff. 40.

Klass, P.J. (3 September 1972). "Keeping the Nuclear Peace: Spies in the Sky," *New York Times Magazine.*

Kohler, R. (2005). "One Officer's Perspective: The Decline of the National Reconnaissance Office." *National Reconnaissance, 2005-U1, 35-44.*

Laurie, C. (2001). *Congress and the National Reconnaissance Office.* Chantilly, VA: National Reconnaissance Office History Staff.

McDonald, R.A. (2002). "Edwin H. Land," in *Recollections of the Pioneers and Founders of National Reconnaissance.* Bethesda, MD: American Society for Photogrammetry and Remote Sensing.

McDonald, R.A. and Moreno, S.K. (2005). *Raising the Periscope; Grab and Poppy: America's Early ELINT Satellites.* Chantilly, VA: Center for the Study of National Reconnaissance.

McElheney, V.K. (1999). "Edwin Herbert Land 1909-1991-A Biographical Memoir." *Biographical Memoirs.* Vol. 7. Washington, D.C.: The National Academy Press.

Morgan, D. and Pincus, W. (5 October 1995). "$1.6 Billion in NRO Kitty Helped Appropriators Fund Pet Projects" *Washington Post.* A15.

Moore, M. (13 June 1991). "Schwarzkopf: War Intelligence Flawed; General Reports to Congress on Desert Storm." *Washington Post.* A1.

Mulcahy, R. D., Jr. (June 2012). *"Corona Star Catchers."* Chantilly, VA: NRO, Center for the Study of National Reconnaissance.

National Reconnaissance Office (November 1994). *National Reconnaissance Office Collocation Construction Project: Joint DoD and CIA Review Report.*

Nowinski, E.H. and Kohler, R.J. (2006). "The Lost Art of Program Management in the Intelligence Community." *Studies in Intelligence.* 50, 2.

Oder, F. C. E., Fitzpatrick, J. C., and Worthman, P. E. (April 2012). *"The Hexagon Story."* Chantilly, VA: NRO, Center for the Study of National Reconnaissance.

Oder, F. C. E., Fitzpatrick, J. C., and Worthman, P. E. (April 2012). *"The Gambit Story."* Chantilly, VA: NRO, Center for the Study of National Reconnaissance.

Outzen, J. D., Ed. (January 2012). *"Critical to National Security: The Gambit and Hexagon Satellite Reconnaissance Systems Compendium."* Chantilly, VA: NRO, Center for the Study of National Reconnaissance.

Outzen, J. D., Ed. (August 2012). *"Trailblazer 1964: The Quill Experimental Radar Imagery Satellite Compendium."* Chantilly, VA: NRO, Center for the Study of National Reconnaissance.

Pedlow, G.W. and Welzenbach, D.E. (1998). *The CIA and the U-2 Program, 1954-1974.* Washington, D.C.: Central Intelligence Agency.

Perry, R.L. (2012). *A History of Satellite Reconnaissance.* Chantilly, VA: Center for the Study of National Reconnaissance.

Pincus, W. (11 August 1994). "CIA Told Hill of Project, Woolsey Says; Others Apologize for Lack of Detail on $310 Million Headquarters," *Washington Post.* A1.

Pincus, W. (12 August 1994). "Spy Agency Defended by House Panel; 'Other Body' Criticized in NRO Building Flap," *Washington Post.* A21.

Pincus, W. (24 September 1995). "Spy Agency Hoards Secret $1 Billion; Satellite Manager Did Not Tell Supervisors of Classified 'Pot of Gold,' Hill Sources Say," *Washington Post.* A1.

Robarge, D. (2007). *Archangel: CIA's Supersonic A-12 Reconnaissance Aircraft.* Washington, D.C.: Central Intelligence Agency.

Ruffner, K.C., Ed. (1995). *Corona: America's First Satellite Program.* Washington, D.C.: Center for the Study of Intelligence, Central Intelligence Agency.

Sapp, B. (21 April 2010). Statement for the Record Before the House Armed Services Committee Subcommittee on Strategic Forces.

Specter, A. (29 September 1995). Floor Statement by the Chairman of the Senate Select Committee on Intelligence, "Intelligence Authorization Act of FY 1996" *Congressional Record.* S14785-6.

Taubman, P. (11 November 2007). "In Death of Spy Satellite Program, Lofty Plans and Unrealistic Bids. *New York Times.* A1.

Thomas, P. (9 August 1994). "Spy Unit's Spending Stuns Hill." *Washington Post.* A1.

van Keuren, D. (1987). "NRL and Space Technology: Peter Wilhelm Reflects on the Origin and Mission of NCST," *NRL Review.*

Weiner, T. (9 August 1994). "Senators Angered Over Cost of Spy Agency's New Offices," *New York Times.* A1.

Weiner, T. (11 August 1994). "Senate Committee Received Apology from Spy Agency," *New York Times.* A1.

Weiner, T. (16 May 1996). "A Spy Agency Admits Accumulating $4 Billion in Secret Money," *New York Times.* A1.

Welzenbach, D.E. (1986) "Observation Balloons and Reconnaissance Satellites," *Studies in Intelligence.* 30, 1.

Appendix A

Timeline of Major Events in NRO History

Date	Event
1949	29 August: Soviet Union tests its first atomic bomb.
1950	4 December: Fearing imminent war with the Soviet Union after the North's invasion of South Korea on 25 June, British Prime Minister Clement Atlee and U.S. President Harry Truman agree to cooperate in overflights of Soviet territory.
1953	8 June: Air Force Scientific Advisory Board (AFSAB) reports that it is possible to build small, lightweight nuclear warheads; this leads to the approval of the Air Force's ballistic missile program, which provides the future launch vehicles for NRO satellites.
1953	12 August: Soviet Union tests a rudimentary hydrogen bomb, which prompts President Truman to establish a "Technical Capabilities Panel" one year later in order to address this new threat. The panel's proposal serves as the foundation for U-2 development.
1954	27 November: President Eisenhower approves the development of the U-2 aircraft.
1955	26 May: President Eisenhower selects NRL to lead the Vanguard research satellite program for the International Geophysical Year.
1956	10 January: Air Force launches the first of 516 high-altitude reconnaissance balloons as part of the Genetrix program to collect imagery of the Soviet Union.
1956	13 January: President Eisenhower establishes the President's Board of Consultants on Foreign Intelligence Activities (PBCFIA), the predecessor of the President's Intelligence Advisory Board.
1956	5 February: Genetrix program conducts last launch; the program ends after Soviet protests.
1956	2 April: WS-117L program office completes the first comprehensive development plan for a reconnaissance satellite.

Date	Event
1956	20 June: First U-2 mission is flown.
1957	3 August: Soviet Union tests the SS-6/R-7, the first ICBM.
1957	4 October: Soviet Union orbits Sputnik 1, first artificial satellite.
1957	24 October: President's Board of Consultants on Foreign Intelligence Activities recommends early consideration of a rush project to develop a space imaging capability.
1957	18 December: President Eisenhower ends the SENSIT airborne reconnaissance program.
1958	31 January: Army launches Explorer-1, the first successful U.S. satellite.
1958	7 February: President Eisenhower approves the Corona program; Corona is split off from WS-117L.
1958	17 March: First successful Vanguard mission.
1958	28 March: Naval Research Laboratory begins development of Grab, the first SIGINT satellite.
1958	29 July: President Eisenhower establishes the National Aeronautics and Space Administration.
1959	21 January: First Corona test fails.
1959	24 August: President Eisenhower gives final approval for NRL to proceed with development of Grab under Project Tattletale.
1960	1 May: CIA U-2 pilot Francis Gary Powers is shot down over the Soviet Union.
1960	22 June: Grab, the first successful reconnaissance satellite, is launched from Cape Canaveral.
1960	18 August: Corona-14 collects the first reconnaissance imagery from space, depicting Mys Shmidta, a Soviet bomber base in northeast Siberia.

Date	Event
1960	25 August: President Eisenhower approves development of the Gambit high resolution IMINT satellite.
1960	31 August: President Eisenhower creates a civilian-led office, the USAF Office of Missile and Satellite Systems, responsible for the Air Force space-based reconnaissance satellite program; this office later serves as the organizational basis of the NRO.
1961	6 September: SECDEF Robert McNamara establishes the National Reconnaissance Office, responsible for managing the National Reconnaissance Program. Joseph V. Charyk and Richard M. Bissell, Jr. are the first NRO co-directors; Bissell serves until 28 February 1962, Charyk serves until 1 March 1963.
1962	25 April: CIA's A-12/Oxcart reconnaissance aircraft makes first flight.
1962	2 May: Second agreement between the Defense Department and CIA organizing the NRO; this agreement provides for a single NRO Director to be appointed by the Secretary of Defense, with the concurrence of the Director of Central Intelligence (DCI).
1962	23 July: Charyk establishes the "alphabet structure" that the NRO will use until 1992: Program A (Air Force satellites), Program B (CIA satellites), Program C (Navy satellites), and Program D (Air Force and CIA reconnaissance aircraft).
1962	13 December: First Poppy mission.
1963	1 March: Brockway McMillan becomes NRO Director; serves until 1 October 1965.
1963	13 March: Third agreement between Defense Department and CIA, making the NRO a separate operating agency with the Defense Department, authorizing the Secretary of Defense to appoint the NRO Director with the concurrence of the DCI, and establishing a NRO Deputy Director to be appointed by the DCI with the concurrence of the Secretary of Defense.
1963	12 July: First Gambit mission.
1963	24 August: First KH-4A (dual film bucket) Corona mission.

Date	Event
1964	29 February: National Security Council approves acknowledging existence of Oxcart; later that day it is announced as the "A-11," an "advanced experimental aircraft" with "military and commercial applications."
1964	21 December: First and only Quill Synthetic Aperture Radar (SAR) experimental satellite launched.
1964	22 December: First flight of SR-71, the Air Force version of Oxcart.
1965	11 August: DCI William Rayburn, Jr. and Deputy Secretary of Defense Cyrus Vance sign agreement removing the requirement for concurrence by the DCI in appointing the NRO Director.
1965	1 October: Alexander Flax becomes NRO Director; serves until 17 March 1969.
1966	President Johnson decides to terminate A-12 program by January 1968; later extended to July.
1966	29 July: First Gambit-3 mission.
1967	15 September: First KH-4B Corona mission.
1968	21 June: Final A-12 flight is completed.
1969	17 March: John McLucas becomes NRO director; serves until 20 December 1973.
1971	22 January: First known mention of the NRO in the press, an article by Benjamin Welles, "Foreign Policy Disquiet Over Intelligence Setup" in New York Times.
1971	15 June: First Hexagon mission.
1971	14 December: Final Poppy mission.
1972	25 May: Final Corona mission.

Date	Event
1973	12 October: First known reference to the NRO in a public record, inadvertently made in a report by the Special Senate Committee to Study Questions Related to Secret and Confidential Government Documents.
1973	21 December: James Plummer becomes NRO Director; serves until 28 June 1976.
1974	1 October: Program D is disbanded; all of its aircraft are transferred to the U.S. Air Force for tactical missions .
1975	3 October: President Ford charters the Civil Applications Committee (CAC), making NRO imagery available to users outside the national security community for the first time.
1976	18 February: President Ford issues EO 11905, "United States Foreign Intelligence Activities," formally establishing the National Foreign Intelligence Program and publicly defining for the first time the U.S. Intelligence Community, including the NRO, which is euphemistically identified with the label "special offices for the collection of specialized intelligence through reconnaissance programs."
1976	28 June: Charles Cook becomes acting NRO Director; serves until 8 August 1976.
1976	9 August: Thomas Reed becomes NRO Director; serves until 7 April 1977.
1977	20 January: First NRO electro-optical satellite system is declared operational by President Carter.
1977	7 April: Charles Cook becomes acting NRO Director; serves until 3 August 1977.
1977	3 August: Hans Mark becomes NRO Director; serves until 8 October 1979.
1978	24 January: President Carter signs EO 12036, "United States Intelligence Activities," giving "full and exclusive authority" over the preparation of the National Foreign Intelligence Program budget (including the NRO) to the DCI.
1978	1 October: To support ratification of SALT II, President Carter confirms that the U.S. possesses and uses intelligence collection satellites.

Date	Event
1978	22 December: Former CIA employee William Kampiles sentenced to 40 years imprisonment for espionage committed in March-April 1977, when he stole a KH-11 manual and sold it to Russian officials in Athens for $3,000.
1979	8 October: Robert Hermann becomes NRO Director; serves until 2 August 1981.
1981	3 August: Edward "Pete" Aldridge becomes NRO Director; serves until 16 December 1988.
1981	13 November: President Reagan signs National Security Decision Directive 8, designating the Space Shuttle as the primary launch system for U.S. government payloads; it directs Defense Department (including NRO) payloads to be compatible with the Shuttle.
1981	4 December: President Reagan signs EO 12333, "United States Intelligence Activities," the current charter (with modifications) under which the U.S. Intelligence Community, including the NRO, operates.
1982	4 July: National Security Decision Directive 42 declares support to deployed military forces as a major space intelligence mission.
1984	17 April: Final Gambit-3 launch.
1985	28 August: A Titan 34-D launched from Vandenberg Air Force Base fails during ascent, resulting in the loss of the payload. This begins a string of mid-1980s launch vehicle problems.
1986	28 January: Loss of Space Shuttle *Challenger* during launch.
1986	18 April: A Titan 34-D fails when a booster explodes shortly after liftoff at Vandenberg Air Force Base, resulting in the loss of the last Hexagon satellite.
1986	1 October: Goldwater-Nichols Defense Reform Act establishes the current statutory process for Defense Department acquisition.
1988	29 September: Space shuttle returns to operation after two years of design review and modifications following the *Challenger* loss.
1988	17 December: Jimmie Hill becomes Acting NRO Director; serves until 27 September 1989.

Date	Event
1989	3 July: The Geiger-Kelly Study, led by RADM Robert Geiger and D. Barry Kelly, recommends a more integrated NRO.
1989	28 September: Martin Faga becomes NRO Director; serves until 5 March 1993.
1992	5 March: The DCI Task Force on the National Reconnaissance Office, chaired by former Lockheed CEO Robert Fuhrman, recommends disbanding the "alphabet structure."
1992	30 March: President George H.W. Bush signs National Security Directive 67, "Intelligence Capabilities 1992-2005," which approves the DCI's recommendation to realign the NRO.
1992	17 September: Government declassifies "fact of" the NRO.
1992	31 December: NRO realignment is put into effect reorganizing along functional lines with directorates for SIGINT, IMINT, and COMM.
1993	6 March: Jimmie Hill becomes Acting NRO Director; serves until 19 May 1994.
1994	19 May: Jeffrey Harris becomes NRO Director; serves until 26 February 1996.
1994	8 August: Sen. Dennis DeConcini (D-Ariz.) Chairman of the Senate Select Committee on Intelligence, and Sen. John Warner, (R-Va.), Vice Chairman, release letter asserting that the NRO concealed the true size and cost of the Westfield HQ, then under construction.
1995	22 February: President Clinton issues Executive Order 12951 declassifying Corona and releasing most Corona imagery to the public.
1995	2 March: President Clinton issued Presidential Decision Directive 35, establishing support to deployed military forces as his top intelligence priority.
1995	24 September: Press reports that the NRO accumulated unused funds from programs. The reports say that DCI John Deutch had responded to complaints from congressional committees in June. Investigations would later place the total "carry over" funds at $3.8 billion.

Date	Event
1996	27 February: DCI John Deutch directs resignation of NRO Director Harris and NRO Deputy Director Jimmie Hill; Keith Hall becomes Acting NRO Director.
1996	26 August: Commission chartered by Acting NRO Director Hall and chaired by retired Adm. David Jeremiah issues its report on the future of the agency.
1996	18 December: First public notification of an NRO launch.
1997	28 March: Keith Hall becomes NRO Director; serves until 13 December 2001.
1997	31 March: Advanced Systems and Technology Directorate (AS&T) is stood up.
1998	17 June: Grab, the first reconnaissance satellite, is declassified in conjunction with the 75th anniversary of the Naval Research Laboratory.
1999	3 September: The NRO announces the award of the Future Imagery Architecture contract to Boeing.
2000	1 November: National Commission for the Review of the National Reconnaissance Office, established by Congress in early 2000, recommends that a portion of the NRO return to its more secretive, compartmented approach.
2001	13 December: Peter Teets becomes NRO Director; serves until 25 March 2005.
2004	11 May: DCI authorizes the NRO to declassify the fact of the Poppy ELINT satellite.
2004	17 December: President George W. Bush signs into law the Intelligence Reform and Terrorism Prevention Act of 2004, reaffirming the NRO as a member of the U.S. Intelligence Community, establishing the Director of National Intelligence (DNI) as head of the Intelligence Community, and transferring the DCI's authorities over the NRO to the DNI.
2005	26 March: Dennis Fitzgerald becomes Acting NRO Director; serves until 25 July 2005.
2005	20 May: NRO retires the BYEMAN control system.

Date	Event
2005	26 July: Donald Kerr becomes NRO Director; serves until 4 October 2007.
2005	28 September: Boeing receives from the NRO partial stop-work order on the Future Imagery Architecture.
2007	19 October: Scott Large becomes NRO Director; serves until 18 April 2009.
2008	1 April: NRO establishes Ground Enterprise Directorate (GED).
2009	12 June: Gen. Bruce Carlson, USAF (Ret) becomes NRO Director, serves until July 2012.
2010	21 September: Director of National Intelligence and Secretary of Defense sign an agreement on the mission, authorities, and responsibilities of the NRO and the NRO Director, superseding the agreement of 11 August 1965.
2011	17 September: DNRO announces declassification of Gambit and Hexagon IMINT satellite systems.
2012	3 July: DNRO announces declassification of programmatic information on the Quill radar satellite system.
2012	6 July: Betty Sapp becomes NRO Director.

Appendix B

Directors of the National Reconnaissance Office

Name	Tour of Duty
Richard M. Bissell, Jr. and Joseph V. Charyk (Co-directors)	Sept 1961 – Apr 1962
Joseph V. Charyk	Apr 1962 – Mar 1963
Brockway McMillan	Mar 1963 – Oct 1965
Alexander H. Flax	Oct 1965 – Mar 1969
John L. McLucas	Mar 1969 – Dec 1973
James W. Plummer	Dec 1973 – June 1976
Charles W. Cook (Acting)	June 1976 – Aug 1976
Thomas C. Reed	Aug 1976 – Apr 1977
Charles W. Cook (Acting)	Apr 1977 – Aug 1977
Hans M. Mark	Aug 1977 – Oct 1979
Robert J. Hermann	Oct 1979 – Aug 1981
Edward C. Aldridge, Jr.	Aug 1981 – Dec 1988
Jimmie D. Hill (Acting)	Dec 1988 – Sept 1989
Martin C. Faga	Sept 1989 – Mar 1993
Jimmie D. Hill (Acting)	Mar 1993 – May 1994
Jeffrey K. Harris	May 1994 – Feb 1996
Keith R. Hall	Feb 1996 – Dec 2001
Peter B. Teets	Dec 2001 – Mar 2005
Dennis D. Fitzgerald (Acting)	Mar 2005 – July 2005
Donald M. Kerr	July 2005 – Oct 2007
Scott F. Large	Oct 2007 – Apr 2009
Betty J. Sapp (Acting)	Apr 2009 – June 2009
Bruce A. Carlson	June 2009 – July 2012
Betty J. Sapp	July 2012 – Present

Index

A

B

C

D

H

Hall, Keith 27, 46, 49
Harris, Jeffrey 25, 45, 46, 49
Hermann, Robert 21, 44, 49
Hexagon 17, 18, 42, 44, 47
Hill, Jimmie 24, 45, 46, 49
House Armed Services Committee 22
House Permanent Select Committee on Intelligence (HPSCI) 24
Hurricane Katrina 32, 34

I

ICBM 7, 9, 20, 40
Ikonos 24
IMINT 1, 7, 8, 9, 11, 19, 23, 40, 45
Integrated Overhead Signals Intelligence Architecture (IOSA) 28
Intelligence Reform and Terrorism Prevention Act of 2004 15, 46
International Geophysical Year (IGY) 6

J

Johnson, Clarence "Kelly" 4

K

Kampiles, William 20, 44
Kelly, Barry 23, 45
Kennedy, President 11
Kennedy Space Center 20
Kerr, Donald 33, 47, 49
Kerrey, Robert 26, 29
KH-1 11
KH-2 11
KH-3 11
KH-4A 11, 42
KH-4B 42
KH-5 18
KH-7 16
KH-8 16
KH-9 17
KH-11 20
Khrushchev, Nikita 19
Killian, James R., Jr. 1, 3, 9

O

Oxcart 4, 41, 42

P

Page, Robert Morris 8
Panetta, Leon 26
Pearl Harbor 2
Perry, William 27
Plummer, James 21, 43, 49
Polaroid 1
Poppy 9, 41, 42, 46
Powers, Francis Gary 4, 11, 40
Presidential Decision Directive 35 22, 45
President's Board of Consultants on Foreign Intelligence Activities (PBCFIA) 9, 39, 40
President's Intelligence Advisory Board 9
Program A 13, 18, 23, 26, 41
Program B 13, 23, 26, 41
Program C 13, 23, 26, 41
Program D 13, 41, 43
Project Aquatone 4

Q

Quill 18, 19, 42, 47

R

RAND Corporation 5, 7
Reagan, President 21, 44
Reed, Thomas 21, 43, 49
RESERVE 33
Ritland, Osmund J. 4, 11

S

SA-2 4
SALT 19, 20, 44
Samos 7, 12
Sapp, Betty 47, 49
Saturn 5
Schriever, Bernard 7, 9
Schwarzkopf, Norman 22
Senate Select Committee on Intelligence (SSCI) 24, 25, 45
SENSINT 3, 40
SIGINT 1, 5, 7, 8, 9, 11, 19, 21, 23, 40, 45
Space Imaging 24

Space Shuttle 5, 21, 22, 44
Space Transportation System 21
Specter, Arlen 25
SPOT 24
Sputnik 1, 7, 11, 19, 40
SR-71 4, 6, 8, 42
Strategic Air Command 9
Strategic Arms Reduction Treaty (START-1) 32
Synthetic Aperture Radar (SAR) 18, 19, 42

T

TALENT KEYHOLE 33
Tattletale 9, 40
Technical Capabilities Panel (TCP) 3, 9, 39
Teets, Peter 33, 46, 49
Thor 7, 9
Titan 7, 17, 21, 22, 44
Truman, President 2, 3, 39

U

U-2 3, 4, 11, 39, 40

V

V-2 4, 5
Vandenberg Air Force Base 21, 44
Vanguard 6, 7, 8, 39, 40
Viking 6, 7

W

Warner, John 24, 45
Westfields 24
Woolsey, R. James 24
WS-117L 7, 9, 11, 39, 40